ENGINEERING TOOLS, TECHNIQUES AND TABLES

FAULT DETECTION: THEORY, METHODS AND SYSTEMS

ENGINEERING TOOLS, TECHNIQUES AND TABLES

Additional books in this series can be found on Nova's website under the Series tab.

Additional E-books in this series can be found on Nova's website under the E-book tab.

ENGINEERING TOOLS, TECHNIQUES AND TABLES

FAULT DETECTION: THEORY, METHODS AND SYSTEMS

LÉA M. SIMON
EDITOR

Nova Science Publishers, Inc.
New York

For permission to use material from this book please contact us:
Telephone 631-231-7269; Fax 631-231-8175
Web Site: http://www.novapublishers.com

LIBRARY OF CONGRESS CATALOGING-IN-PUBLICATION DATA

Fault detection : theory, methods, and systems / editor, Lia M. Simon.
p. cm.
Includes bibliographical references and index.
ISBN 978-1-61728-291-1 (hardcover)
1. Fault location (Engineering) 2. Defects. I. Simon, Lia M.
TA169.6.F375 2010
620'.0044--dc22

2010015590

Published by Nova Science Publishers, Inc. † New York

CONTENTS

PREFACE

Fault detection is a subfield of control engineering which concerns itself with monitoring a system, identifying when a fault has occurred and pinpointing the type of fault and its location. Two approaches can be distinguished: a direct pattern recognition of sensor readings that indicate a fault and an analysis of the discrepancy between the sensor readings and expected values. This book reviews research in the field of fault detection including the introduction of an automatic, effective but simpler-to-use system for fault diagnosis in bearings; the monitoring and diagnosis of discrete event systems using time petri nets; fault detection and diagnosis using statistical and soft computing methods.

Maintenance is essential in all kinds of machinery. In order to prevent long-term breakdown or catastrophic failure, faults should be detected at their incipient stage. Bearing faults are the most frequent cause of failures in machinery. Most of the faults occurring in bearings will introduce impact vibration signals when bearings are rotating. Such impacts behave like damped oscillations and can be detectable by using accelerometers. In the past, maintenance staff assessed a machines' condition by measuring the increase of either the overall vibration level or the magnitudes of characteristic frequencies generated by roller bearings. To date, many staffs still adopt these methods for vibration-based machine fault diagnosis. However, these methods are not effective. Bearing vibrations displayed in a frequency spectrum are often overwhelmed by higher vibrations generated by larger components located adjacent to the inspected bearing. Faulty bearing information could be difficult to reveal. Hence, maintenance staff fail to detect bearing faults at their earlier stages prior to a fatal breakdown of bearing.

Research has been conducted to invent effective techniques so that such anomalous impacts can be more easily detected either in its time or frequency domains. For example, the ever-popular Fast Fourier transform, the Wavelet transform, the Hilbert-Huang transform etc. are aimed to detect bearing faults through the use of either conventional frequency spectrum or advanced decomposed waveforms that may exhibit instantaneous changes in temporal and frequency characteristics. Meanwhile, while these techniques have proven to be successful in various degrees of fault detection, the industry is reluctant to adopt these techniques as they require intensive analysis and experts to interpret the results.

The purpose of Chapter 1 is to introduce an automatic and effective but simpler-to-use system for bearing fault diagnosis that the industry can apply to its equipment without the need of hiring experts. The system uses the reassignment wavelet transform to decompose the captured vibration signals into a number of waveforms at different frequency bands. Based on the calculated values of kurtosis and root-mean-square (RMS) for each decomposed band, the system will automatically select the most suitable frequency band for detecting whether impact occurs. If the appearance of impacts is confirmed, then the system will automatically extract the characteristics of impacts for further fault classification.

The effectiveness of the system was verified by both simulated data and real machine generated vibrations. Moreover, a comparison study was conducted to show the results generated by several popular techniques used in this system. The system has been designed and implemented as a series of virtual instruments with man-machine interface. The process of developing the system as well as the required data acquisition and sensors are briefly described in the chapter. By incorporating this system in bearing maintenance practice, the industry may enjoy not only the convenience in automating the process of bearing fault detection, but also the enhanced accuracy in determining the possibility of bearing failure which may eventually lead to the complete breakdown of the inspected machine.

Monitoring of discrete event systems has been the subject of several studies during the last decades. The approaches developed are different as regards the model used, the hypothesis taken and the point of view considered. In Chapter 2, a monitoring approach for timed discrete event systems is discussed. The system behavior is modeled using Time Petri Net. The proposed approach exploits temporal constraints to assess the system state and therefore detect and identify faults being given the partial observability of events. This approach uses an offline synthesis of the timed system behavior which serves as a basis to the online monitoring. Then, an online algorithm is used to track the system's state and to identify the event scenarios which occur within the system. Several mechanisms have been elaborated in order to detect the faulty behaviors and to identify the possible failures. They are based both on the structural analysis of the system behavior and on the time constraints. Moreover, the developed approach offers the ability to anticipate the occurrence of some failures which is an interesting fact, especially when dealing with critical systems. Diagnosability questions are also put forth and some results are proposed. A railway control case study is used in order to gradually illustrate the various approach steps.

Most of the research work that has been done on the field of fault detection and diagnosis has used the model based approach. Chapter 3 presents an alternative way of carrying out a complete fault detection and diagnosis system using statistical and soft computing methods. This proposal is based only on the system's or process' history data treatment. The motivation of using process history data is basically: 1. To obtain an approach that could take into account variables correlations, that at first sight could not be so clear even for expert designers, 2. To develop a diagnostic system that learns normal operation mode directly from the process, instead of having a model based diagnostic system which depends on the expertise of the designer to manage the complexity of the system when modeling. The advantage of having a process history data based approach over a model based framework is the relatively easy way to obtain data from automated industrial processes. In most of the modern systems, can be very difficult to obtain an exact model due to the big quantity of information needed and the variables correlations, which can cause false alarms, indicating a wrong faulty component or system. Nevertheless, this kind of approaches combining statistical and soft computing methods can be supported or complemented with model based methods, in order to have a more powerful diagnosis method combining the expertise and mastery of the designer and those hidden behaviors that many systems exhibit. In this chapter it is shown how statistical methods applied in a straightforward way and combined with soft computing methods such as artificial neural networks and fuzzy logic, are ideal tools for doing diagnosis. This translates to finding the root cause of the problem using only process history data as a prior knowledge of a system, no matter if it is linear or nonlinear. This knowledge is used to give a final diagnosis, in complex scenarios whith noise presence and correlated variables that could easily mislead to false alarms or wrong diagnosis. The organization of this chapter is as follows. First of all it is presented an introduction of why a fault diagnosis is necessary. Then a classification of fault diagnosis methods is given. After that, a presentation of the mathematical tools used is shown and then how they have been tailored in the authors' research, in order to build complete fault detection and diagnosis systems for several applications. The authors present case studies that show promising results using the algorithms proposed. Finally the conclusion over this chapter is given.

Chapter 4 addresses Gross Error Detection using uni-variate signal-based approaches and an algorithm for the peak noise level determination in measured signals. Gross Error Detection and Replacement (GEDR) may be carried out as a pre-processing step for various model-based or statistical methods. More specifically, this work presents developed algorithms and results using two uni-variate, signal-based approaches regarding performance, parameterization, commissioning, and on-line applicability. One approach is based on theMedian Absolute Deviation (MAD) whereas the other algorithm is based on wavelets. In addition, an algorithm, which was developed for the parameterization of the MAD algorithm, is also utilized to determine an initial variance (or peak noise level) estimate of measured variables for other model-based or statistical methods. The MAD algorithm uses a wavelet approach to set the variance of the noise in order to initialize the algorithm. The findings and accomplishments of this investigation are:

1. Both GEDR algorithms, MAD based and wavelet based, show good robustness and sensitivity with respect to one type of Gross Errors (GEs), namely outliers.
2. The MAD based GEDR algorithm, however, performs better with respect to both robustness and sensitivity.
3. The algorithm developed to detect the peak noise level is accurate for a wide range of S/N ratios in the presence of outliers.
4. The two developed algorithms based on wavelets (Algorithms for GEDR and peak noise level estimation) do not require the specification of any additional parameters which makes parameterization fully automated as a result. The MAD algorithm only requires the result of the peak noise level detection algorithm that has been developed and therefore is also fully parameterized.
5. The sensitivity and robustness of the algorithms is demonstrated using computer-generated as well as experimental data.
6. The algorithms work both, for off and on-line cases where in the online case computation times are small so that application on standard Distributed Control Systems (DCS) for large plants should be feasible.

7. A translation fromMatlab to C (C++) was done using theMatlab Compiler and a COM wrapper was implemented as a functional prototype, which can be directly accessed through an excel spreadsheet.

To conclude, the developed algorithms are totally general and they are present in some industrial software platforms to detect sensor outliers. Furthermore, it is currently integrated in the inferential modelling platform of the unit responsible for Advanced Control and Simulation Solutions within ABB's (Asea Brown Boveri) industry division. Experimental results using sensor measurements of temperature, pressure and Enthalpy in a Distillation Column are presented in the paper.

In Chapter 5 the authors present a technique which generates from Abstract State Machines specifications a set of test sequences capable to uncover specific fault classes. The notion of *test goal* is introduced as a state predicate denoting the detection condition for a particular fault. Tests are generated by forcing a model checker to produce counter examples which cover the test goals. The authors introduce a technique for the evaluation of the fault detection capability of a test set. The authors report some experimental results which validate the method, compare the fault adequacy criteria with some classical structural and combinatorial coverage criteria and show an empirical cross coverage among faults.

Electromechanical components manufacturing companies more often ask for automatic on-line inspection systems in order to accurately monitor the characteristics of all their products and components.

Condition monitoring of manufacturing appliances is often based on the analysis of machines' vibrations, as the emergence of the fault can be indicated by the variation of the vibration signals that they produce. This is because when a machine or a structural component is in good condition, its vibration profile has the "normal" characteristic shape, and it will change as a fault begins to develop.

Piezo-electric accelerometers and microphones are the most common vibration sensors used in fault diagnostics. However, they appear to be less attractive for on-line quality control because of several limitations, such as invasiveness, problematic installation and high sensitivity to background noise.

For these reasons, the use of Laser Doppler Velocimeters (LDV) has become an increasingly popular technique owing to the non-contact principle of the laser. However, despite the advantages of LDV, vibration measurements on rough surfaces can be distorted by undesired surface effects, such as the speckle noise.

Acquired vibration signals have to be processed and analysed in order to identify the characteristics that allow distinguishing between the good and the faulty machinery. To this end, several techniques of signal processing have been developed in the last years such as FFT based analysis, Wavelet analysis, Order analysis, etc.

After a brief introduction on the vibration theory for on-line diagnosis, Chapter 6 presents an overview of the most common vibration measurement transducers and data analysis techniques, followed by some examples of their applications in the development of test benches for industrial production lines both on electro-mechanical components, such as universal motors and on assembled complex systems, such as washing machines.

In: Fault Detection: Theory, Methods and Systems ISBN: 978-1-61728-291-1
Editor: Léa M. Simon, pp. 1-67

Chapter 1

ADVANCED SYSTEM FOR AUTOMATICALLY DETECTING FAULTS OCCURRING IN BEARINGS

W. Peter Tse * ***and T. Jacko Leung***
Smart Engineering Asset Management Laboratory (SEAM)
Department of Manufacturing Engineering & Engineering Management
City University of Hong Kong, Hong Kong, China

Abstract

Maintenance is essential in all kinds of machinery. In order to prevent long-term breakdown or catastrophic failure, faults should be detected at their incipient stage. Bearing faults are the most frequent cause of failures in machinery. Most of the faults occurring in bearings will introduce impact vibration signals when bearings are rotating. Such impacts behave like damped oscillations and can be detectable by using accelerometers. In the past, maintenance staff assessed a machines' condition by measuring the increase of either the overall vibration level or the magnitudes of characteristic frequencies generated by roller bearings. To date, many staffs still adopt these methods for vibration-based machine fault diagnosis. However, these methods are not effective. Bearing vibrations displayed in a frequency spectrum are often overwhelmed by higher vibrations generated by larger components located adjacent to the inspected bearing. Faulty bearing information could be difficult to reveal. Hence, maintenance staff fail to detect bearing faults at their earlier stages prior to a fatal breakdown of bearing.

* E-mail address: mepwtse@cityu.edu.hk. (Corresponding author)

Research has been conducted to invent effective techniques so that such anomalous impacts can be more easily detected either in its time or frequency domains. For example, the ever-popular Fast Fourier transform, the Wavelet transform, the Hilbert-Huang transform etc. are aimed to detect bearing faults through the use of either conventional frequency spectrum or advanced decomposed waveforms that may exhibit instantaneous changes in temporal and frequency characteristics. Meanwhile, while these techniques have proven to be successful in various degrees of fault detection, the industry is reluctant to adopt these techniques as they require intensive analysis and experts to interpret the results.

The purpose of this chapter is to introduce an automatic and effective but simpler-to-use system for bearing fault diagnosis that the industry can apply to its equipment without the need of hiring experts. The system uses the reassignment wavelet transform to decompose the captured vibration signals into a number of waveforms at different frequency bands. Based on the calculated values of kurtosis and root-mean-square (RMS) for each decomposed band, the system will automatically select the most suitable frequency band for detecting whether impact occurs. If the appearance of impacts is confirmed, then the system will automatically extract the characteristics of impacts for further fault classification.

The effectiveness of the system was verified by both simulated data and real machine generated vibrations. Moreover, a comparison study was conducted to show the results generated by several popular techniques used in this system. The system has been designed and implemented as a series of virtual instruments with man-machine interface. The process of developing the system as well as the required data acquisition and sensors are briefly described in the chapter. By incorporating this system in bearing maintenance practice, the industry may enjoy not only the convenience in automating the process of bearing fault detection, but also the enhanced accuracy in determining the possibility of bearing failure which may eventually lead to the complete breakdown of the inspected machine.

Section 1. Introduction

Much work has been done on how to detect fault-development indicators that point to machine breakdowns. Bearing faults are the most frequent causes of failures in rotating machinery. However, signals generated by bearings are often overwhelmed by higher amplitude signals from nearby components or surrounding noise. Useful information is thus lost and the maintenance personnel can hardly perform proper remedies to defective machines.

In traditional approaches, Wavelet is one of the commonly adopted tools for analyzing the faulty bearing signal. However, Conventional Wavelet is limited by its energy leakage. To solve this shortcoming, Reassignment Wavelet is introduced in this chapter to concentrate and localize the defective bearing signal.

One of the aims of this chapter is to use Reassignment method to minimize the energy leakage caused by Conventional Wavelet. Also, a Spectrum of Root Mean Square (RMS) and Kurtosis-based method is introduced to automatically detect the location of the bearing excitation zone. Finally, through the decomposition analysis, the impact intervals in time can be revealed for determining the kind of bearing faults.

In this chapter, a completely new and automatic fault detection system with a well-established virtual instrument will be introduced. Comprehensive data was collected from both a laboratory fault simulating machine and industrial machineries. In addition, a modified method, Reassignment Wavelet based Spectral RMS x Kurtosis, has been developed to facilitate the identification of anomalous impacts appearing in different defective machineries. The results successfully reveal that this method is indicative in detecting the bearing faults at early stage. In particular, with the help of Reassignment Wavelet decomposition and Spectral RMS x Kurtosis, both naturally developed and artificially induced impacts of the defective bearings could be located even though it is presented in a noisy environment. Moreover, by adopting this automatic fault detection system, fault detection on machinery become easy and it can save maintenance costs by preventing the need to hire experts.

1.1. The Background of Machinery Fault Detection

Maintenance is essential in all kinds of machines. In the past, machine operators would recognize the machine's condition by touching the machine or measuring the overall vibration level with a digital vibration meter. However, this is too subjective and not effective. According to Kiral [0], it is important to detect a defect at its incipient stage in order to prevent long-term breakdowns or in some cases possible catastrophic failure.

In fact, most modern machineries are so complex that many components may run together, making it impossible for the operator to distinguish between a normal and anomalous machine. Fast Fourier Transform (FFT), has been applied in machine monitoring to discriminate the machine vibration signature, however, FFT is not sufficient to detect instantaneous changes due to its lack of time representation. According to Peng 0, FFT-based analyses are not suitable for non-stationary signals and are not able to reveal the inherent information of non stationary signals. Therefore, an alternative to FFT, called Wavelet Transform, is investigated.

Definitely, when machines start to become defective, the force acting on different components will be changed, which will result in changes in the vibration spectrum. Therefore an innovative technique, namely the Continuous Wavelet Transform (CWT), was recently introduced. CWT is an innovative technique, which combines both the time and frequency analysis by shifting the well defined mother Wavelet from different scales. According to Tse 0, CWT is very useful in detecting both periodically and randomly occurring anomalous signals. Also, CWT compensates for the deficit from FFT. This is because CWT decomposes the original vibration signal into different frequency sub-bands and provides a time representation for individual analysis. Therefore, CWT was widely adopted for vibration based machinery fault diagnosis. Unfortunately, the conventional CWT also has its own weakness which is energy leakage 0. According to Tse, Yang and Tam 0, the overlapping problem under Wavelet analysis would cause a large amount of redundant information in the results. This could smear the Spectral features. Therefore, there is a need for novel analysis of machinery diagnosis, which will solve the overlapping issue and is capable of detecting impulsive signals caused by defective components. To solve the problem, this chapter aims to investigate the Reassignment method. Finally, the spectrum of RMS and Kurtosis is generated, based on the Reassignment Wavelet, which helps to determinate the most suitable frequency band for bearing fault representation.

1.2. The Motivation to Develop an Automatic System for Detecting Bearing Faults

Maintenance is one of the serious problems faced by many industries and utilities. According to Neale 0 the purchasing and installation costs of the equipment usually cost less than half of the total expenditure spent over the life of the machine. According to Wowk 0, maintenance expenditure typically represents 15 to 40 percent of the total cost and it can be up to 80 percent of the total expenditure.

At the beginning of this chapter, it has been mentioned that lots of research has been conducted to invent effective techniques on detecting bearing fault and these techniques have been proved to be a success in various degrees of fault detection. However, the industry is still reluctant to adopt these techniques as they require intensive analysis and experts to interpret the results. Hence, the purpose of this chapter is to introduce an automatic, low cost, effective and simpler-to-use system for bearing fault diagnosis that the industry can apply to its equipment without the need of hiring experts.

Besides the development of such automatic fault detection system, all the virtual instruments used were enhanced with several innovative signal processing techniques so that the captured symptoms could be identified in different machinery. To ensure that the fault could be detected as early as possible, appropriate statistical analyses were implemented to retread the faulty signal for different machinery. Hence, the effectiveness for machinery fault detection could be significantly improved.

In this chapter, a novel system will be developed for bearing fault detection. It will be embedded with all the aforementioned technologies and introduce an advanced method, namely, "Reassignment Wavelet based Spectral RMS x Kurtosis", which can automatically detect the location of the bearing excitation zone. Due to the limit of the page length, this chapter has concentrated on examining bearings only. However, the proposed methodology can also be applied to different kinds of mechanical components, such as pumps or gears.

1.3. The Review of the Maintenance Approaches

Machine maintenance can be classified into three approaches. One is the "Run-to-Breakdown and Replace approach", the second is the "Time-Based maintenance approach" and the last is the "Condition-Based maintenance approach".

1.3.1. The Run-to-Breakdown Approach

In industry, most machines run until breakdown. Loss of production is not significant as a spare machine or other production line will usually take over. In this situation, knowing the machine's condition is not advantageous as there is usually no economic or safety advantage in knowing when a breakdown will occur, as illustrated in Figure 1. However, for public utilities, machine malfunction would cause a huge profit loss in the economy. It may not be possible to perform maintenance after a machine has broken down.

1.3.2. The Time-Based Preventive Maintenance Approach

In this approach, maintenance operations are often performed at fixed intervals such as every 3,000 operating hours or once per year. It is normally determined statistically by the time when the machines are in a fully serviced condition. This approach normally depends on the history of the machine's condition. Figure 2 shows the concept of the time based maintenance approach.

The maintenance process is performed after a constant period, which is in order to maintain the operating condition.

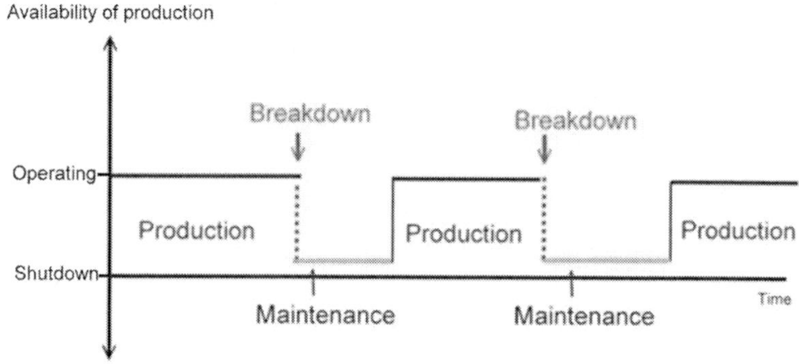

Figure 1. A conceptual diagram of the run-to-breakdown maintenance approach.

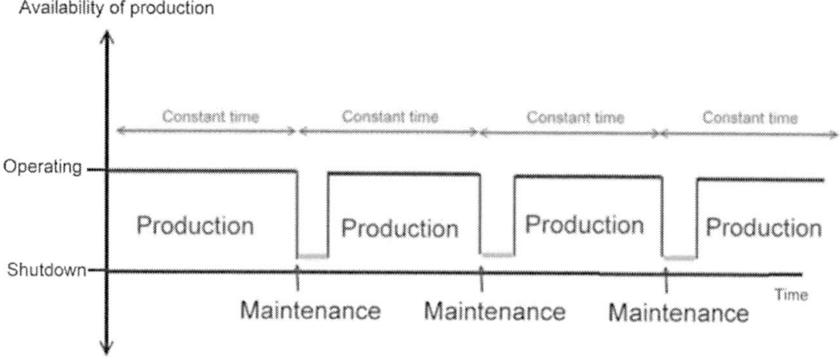

Figure 2. A conceptual diagram of the time-based maintenance approach.

1.3.3. The Condition-Based Maintenance Approach

Condition-based maintenance (CBM) would consider each machine individually and provide a low cost but more effective approach for the maintenance engineer. Figure 3 illustrates the concept of the condition-based maintenance approach. In this approach, the maintenance schedule is based on the machine's condition, such as the vibration level. The cost of the machine maintenance is reduced, since the maintenance resources are rearranged more effectively to perform the maintenance schedule.

According to Figure 3, the maintenance schedule of condition-based maintenance is planned based on the vibration level of the machinery, i.e. based on the machine's health. According to ISO 10816 0, different type of machines could be broken-down when the overall vibrations have exceeded the pre-defined levels. Unlike the time-based maintenance approach, CBM will consider the machine condition when it is used for maintenance service. It means that maintenance can always be done before the machine's vibration level has reached an unacceptable levels, which defined by ISO 10816 0.

1.3.4. The Advantages for Adopting the Conditional-based Maintenance Approach

The advantages of adopting the condition-based maintenance approach are as follows:

1. Prevents sudden breakdown of machines and possible losses due to accidents
2. Makes the best use of components. In other words, lowers the wastage in replacing components earlier than is necessary
3. Reduces the expertise required for data collection and analysis.

Although time-based preventive maintenance can prevent sudden breakdown, it is not efficient because maintenance would be scheduled at a constant period, e.g. overhaul, even though the machine is operating in good condition. In this situation, part of the resources and labor would be wasted because maintenance is not scheduled efficiently. Furthermore, the past maintenance history shows that in the vast majority of cases, time-based preventive maintenance is not economical. As the actual failure pattern for each individual machine cannot be predicted, time-based preventive maintenance may not be applied. Therefore, a more efficient approach is needed which is the condition-based machine maintenance.

At present, lots of industries are employing both the run-to-failure and preventive maintenance approaches. Although there are comprehensive preventive maintenance approaches in local industry, huge catastrophes still cannot be totally avoided due to the probabilistic nature of machine failure. Mechanical failure during our daily lives, such as a breakdown in an air ventilation system, or the failure of a car exhaust engine, could result in a long repair and recovery time. In that case, a company would suffer serious complaints from customers or passengers. The government is therefore very concerned with maintenance strategy in view of the fact that maintenance reliability is so important to industry.

Therefore, the requirements for the maintenance system become critical to the success of the industry.

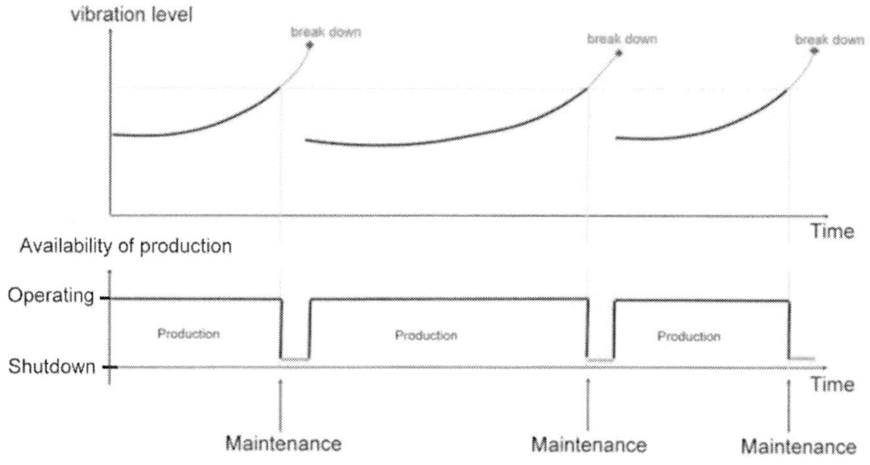

Figure 3. A conceptual diagram of condition-based maintenance.

As mentioned in the introduction, vibration is one of the most effective indicators for a machine's condition. This is because a repetitive impact vibration signal is normally a destructive by-product of the force transmission through a machine. Based on the existence of impact, the operator can predict the machine break down period. As a result, the cost of maintenance is reduced.

Section 2. The Review of the Machinery Fault Detection Methodology

Vibration signal analysis has long been used for machinery fault detection 0 0. The general principle is based on the analysis of signals that are issued from components in mechanical systems during operation. When faults are developed in a system, some of the system dynamics vary, resulting in significant deviations in the vibration pattern. This dissertation focuses on vibration signals and it can be collected by using transducer. Professionals usually study the sudden changes of vibration signals from different frequency ranges because this often implies breakdowns or early defects of specific components in the machinery.

Machine elements react against each other and energy is dissipated through the structure in the form of vibration. Therefore, vibration signals carry much information relating to the operating conditions. As mentioned in previous sections, vibration is normally a destructive byproduct of the transmission force through a machine which provokes wear and accelerates breakdown. For most machines, the vibration has a typical level and its frequency spectrum has a characteristic shape when the machine is in good condition. A plot of the vibration spectrum is known as the vibration signature of the machine, it can be trended and analysed by different signal processing techniques, such as the Wavelet Decomposition and Spectral Kurtosis techniques.

According to Sadettin 0, vibration analysis is a technique which is being used to track machine operating conditions and trend deteriorations in order to reduce maintenance costs and downtime simultaneously. It can basically derive into two parts; one is the data collection by means of the vibration analyser equipped with transducer, the other is to prevent catastrophic failure by detecting the impact caused by defective components with any signal processing technique.

2.1. Transducer Used for Measuring Vibration Signal

The transducer used to measure the vibration signal is called an Accelerometer. It is a seismic instrument and is used to sense the vibration signals generated by the inspected components. It is made of a small mass of piezoelectric crystals. The crystals act as a stiff spring as well as generating a column of charge proportional to acceleration, as illustrated in Figure 4. The design of the accelerometer was brought through a cable to a voltage amplifier. The amplifier will convert the vibration to voltage, which is proportional to the acceleration.

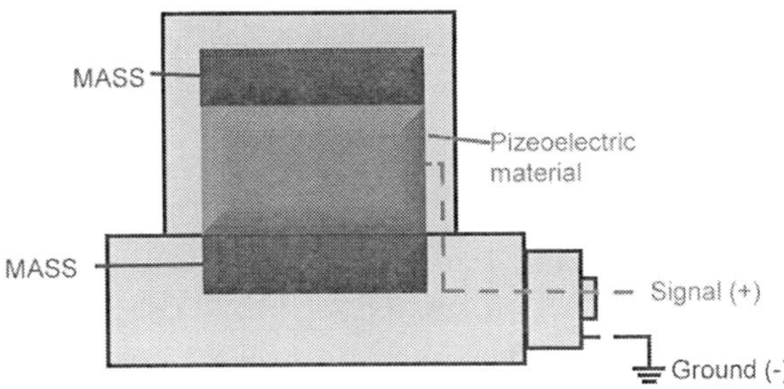

Figure 4. The integrated-circuit piezoelectric (ICP) accelerometer.

2.2. Vibration Measurement

Vibration phase and amplitude can be detected via a piezoelectric (PZT) accelerometer. PZT accelerometers rely on the piezoelectric effect of quartz or ceramic crystals to generate an electrical output that is proportional to the applied acceleration. The piezoelectric effect produces an opposed accumulation of charged particles on the crystal. This charge is proportional to the applied force or stress. Therefore, the amplitude of vibration can be detected through the detection of the charge unit. A force applied to a quartz crystal lattice structure alters the alignment of positive and negative ions, which results in an accumulation of these charged ions on opposed surfaces. These charged ions accumulate on an electrode that is ultimately conditioned by transistor microelectronics.

In an accelerometer, the stress on the crystals occurs as a result of the seismic mass imposing a force on the crystal. Over its specified frequency range, this structure approximately obeys Newton's law of motion, as illustrated in Equation (1).

$$F = ma. \quad (1)$$

Where '*F*' is Force, '*m*' is mass and '*a*' is acceleration.

Therefore, the total amount of accumulated charge is proportional to the applied force, and the applied force is proportional to the acceleration. Electrodes collect and wires transmit the charge to a signal conditioner that may be remote or built into the accelerometer. Once the charge is conditioned by the signal conditioning electronics, the signal is available for display, recording, analysis, or control. Therefore, the machinery fault symptom can be detected via the detection of the output signal from the PZT accelerometer.

2.3. Vibration Trend

The waterfall plot function for vibration trend analysis is widely adopted in machinery fault detection. The vibration trend is a 3D plot that consists of many spectra obtained at different times or days. As shown in Figure 5, the y-axis of the trend plot is the vibration amplitude, the x-axis is the frequency, and the z-axis is the number of records obtained at different times. Since the trend plot shows the temporal information of the vibration, it is helpful to observe the change of vibration in a particular range of frequencies. If a component in a machine is deteriorating, its vibration level will be continuously increasing.

Hence, by observing the increase in vibration level in the trend plot, one can determine the seriousness of a damaged component. However, this technique requires the existence of a set of normal operating conditions, which are usually named as 'baseline' by the maintenance people. Also, a good record or history track of the machine operating condition is necessary when we are adopting this trending technique. However, as previously mentioned that lots of industries still adopting the time base preventive maintenance approach, they may not available to provide the useful operating history for diagnosis of their machinery. Therefore, this technique cannot always successfully detect the fault before the machine breaks down.

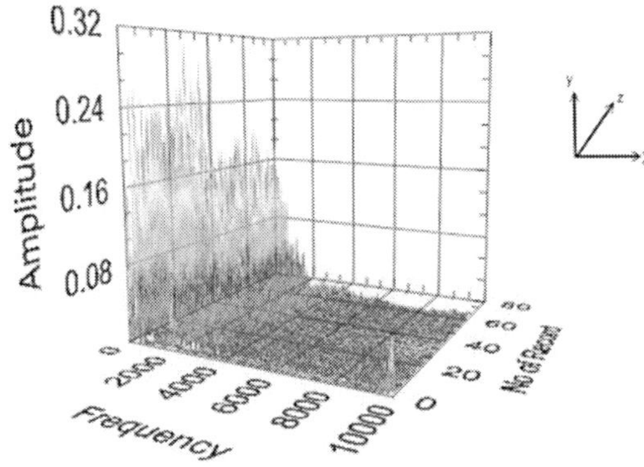

Figure 5. The vibration trend analysis embedded in my system.

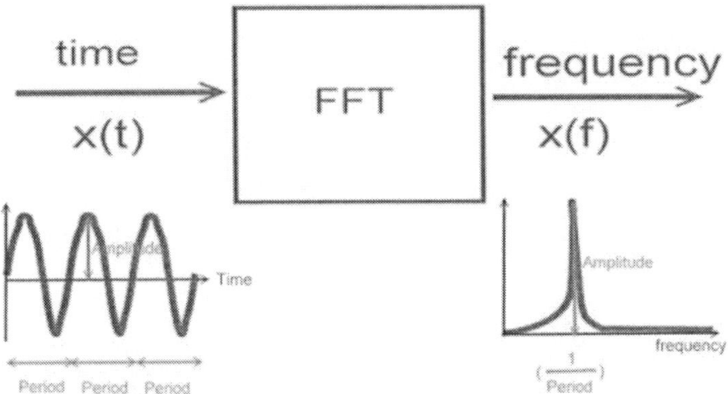

Figure 6. An example of FFT.

2.4. The Introduction of Fast Fourier Transform (FFT)

One of the most common methods in vibration analysis is called the Fast Fourier Transform (FFT). It transforms a time domain signal into a frequency domain with different amplitudes corresponding to the bandwidths. For example, when applying FFT to a sine signal with a known frequency, the corresponding peak should be found on the spectrum result of the FFT, as illustrated in Figure 6.

According to Wowk 0, the frequency is the characteristic of the source and the amplitude is the characteristic of the path. The FFT spectrum shows the magnitude of the signal in the frequency domain. If the bearing has defects, the impact signal should be generated by the defective bearing and this would produce abnormal components in the spectrum. Since abnormal vibration impact would always produce a high frequency, the operator could check the vibration spectrum, such as the FFT plot, in order to indicate the machine's condition.

2.4.1. The Principal of Fast Fourier Transform

In the previous section, it was mentioned that FFT is a popular and widely adopted tool for vibration based machine fault diagnosis. It constructs a frequency spectrum from the captured signal. A typical vibration signal is shown at the top of Figure 7, its corresponding FFT spectrum is shown below. In Figure 7, we can see that the collected signal contains a vibration signal generated from a different source. By using the FFT, the vibration signal can mainly divide into two frequency ranges, one exists from 200 Hz to 300 Hz, and the other exists at around 1000 Hz.

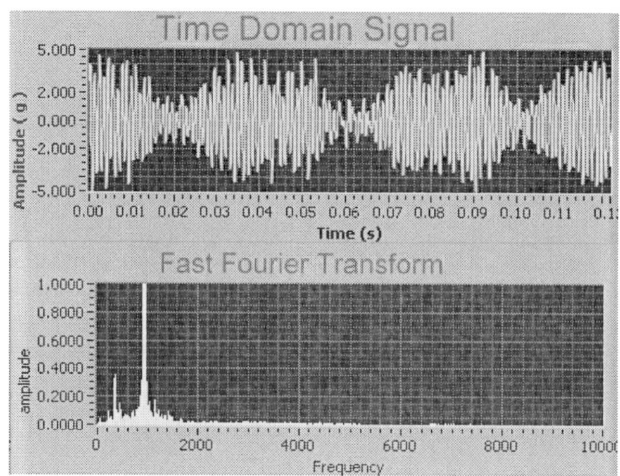

Figure 7. The spectrum shows the magnitudes of signals in the frequency domain.

Definition of FFT

The definition of FFT shows that a periodic waveform can be represented as the sum of a fundamental plus its harmonics as defined in Equation (2)

$$x(t) = \frac{a_0}{2} + \sum_{n=0}^{\infty}[a_n \cos(n\omega t) + b_n \sin(n\omega t)] \quad (2)$$

In Equation (2), $\frac{a_0}{2}$ is represented as the zero frequency amplitude, or it is the offset value of the signal, where,

$$a_0 = \frac{2}{T}\int_{-T/2}^{T/2} x(t)\delta t,\ a_n = \frac{2}{T}\int_{-T/2}^{T/2} x(t)\cos(n\,\omega t)\delta t,$$

$$b_n = \frac{2}{T}\int_{-T/2}^{T/2} x(t)\sin(n\,\omega t)\delta t$$

In this form, the series is from negative infinite to positive infinity including the zero term. It may be the most difficult form to be calculated. However, it may be the most useful form of representation, especially in the frequency domain of electronic circuit analysis.

According to Tse and Lai 0, if a component of the machine is defective, the magnitude of the vibration generated from this component will increase at its related excitation frequency. Comparisons can be made to confirm the existence of faults by using the spectrum generated in the normal operating conditions.

2.4.2. The Limitations of the Existing FFT

Fourier proved that any continuous periodic function *f(t)* with period(T), can be divided as a set of sinusoids and cosine components with separated correspond frequent 0. However, the limitation for Fourier Transform is that it functions with the periodic signal only. Fourier Transform may not function as well for transient signals or instantaneous information. Hence, Wavelet Transform was investigated to examine the time dependent variations of instantaneous frequency contents.

2.5. The Introduction of Wavelet Transform

As mentioned before, machinery fault diagnosis is one of the major considerations in industry, and vibration analysis has been widely used for machinery fault diagnosis. The key issue is how to extract the useful features from these vibration signals. In the previous section, FFT was introduced, which uses sinusoidal functions as basic functions for extracting the frequency information and it is one of the well-established methods in vibration analyses. However, FFT analysis provides zero representation of signals localised in time. For this reason, a time-frequency analysis tool is necessary to localise the signal in both time and frequency domains. Wavelet is one of the effective tools for this. It provides the characteristics of both the time and frequency domains, which gives the additional information of time series over the FFT.

2.5.1. The Principal of Wavelet Transform

Wavelets divide data into different frequency sub-bands so that each sub-band may be analysed individually. Therefore, they are advantageous over traditional FFT methods in analysing vibration signals. A typical type of Wavelet Transform is an Integral Transform, defined in Equation (3), in which a signal is represented in terms of a family of time and frequency localised basic functions.

According to Chui 0, the Wavelet Transform, $W_{(a,b)}$, can be expressed as Equation (3)

$$W_{(a,b)} = \frac{1}{\sqrt{a}} \int_{-\infty}^{\infty} f(t)\, \varphi^* \left(\frac{t-b}{a}\right) dt \qquad (3)$$

where, *a* is the scale, *b* is the translation value, *φ(t)* the mother Wavelet, (*) represents the complex conjugate and *f(t) is a* continuous-time function.

The main purpose of the mother Wavelet is to provide a source function to generate the daughter Wavelets which are simply the translated and scaled versions of the mother Wavelet. In this research, Morlet was selected as the mother Wavelet. This is because the base of Morlet is an exponentially decaying cosine type of sinusoidal function. Also, the decade function is significantly similar with the decaying part of the typical impact, as illustrated in Figure 8. Besides, the impact can be identified and magnified in each frequency scale by adjustment of the exponential value (decay rate) in Morlet. Therefore, this makes Morlet Wavelet the suitable mother Wavelet function in this research.

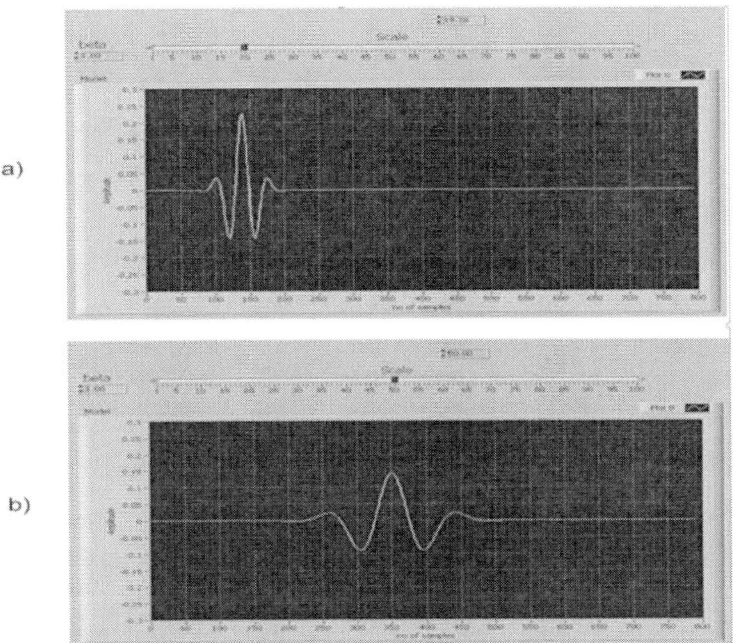

Figure 8. (a) A Morlet function in scale = 19.78. (b) The same Morlet function in scale = 50.

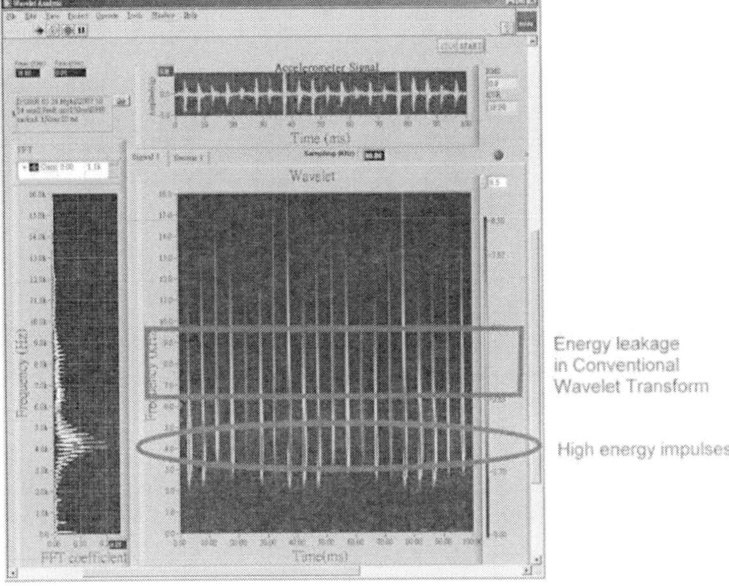

Figure 9. Energy leakage in conventional Wavelet Transform.

A machine integrated with various kinds of components should have different excitation frequencies. Hence, the vibration generated by each component can be revealed at its related excitation frequency. By observing the change of each excitation frequency, the health of the related component can be determined.

2.5.2. The Limitations of Existing Wavelet Transform

One of the detected vibration signals, which was analysed by using conventional Wavelet Transform, is shown in Figure 9. As indicated, the bright colour spots, as highlighted with a red circle, reveal the high-energy impulses caused by anomalous vibrations. However, the leakage phenomenon exists, as highlighted with red squares, which makes the determination of the excitation frequency zone very difficult.

According to Equation (3), the Wavelet coefficients taken at every point are the summarised values which would be assigned to the geometric centre of the entire energy distribution. However, the energy distribution is not always geometrically symmetric. Assigning the Wavelet coefficient to this geometric centre may not be sufficient to represent the entire energy distribution. Therefore, the locations of the coefficients are not correctly coordinated. As a result, energy may leak from the geometrical centre due to this incorrect coordination. In serious cases, such shortcomings will interrupt the identification of the excitation frequency layer or miss the true fault symptoms. Therefore, a more sophisticated technique is needed to solve the leakage problem and improve the readability of the impacts. This is the Reassignment Wavelet. Reassignment is a technique that overcomes the trade-off between localisation and interference in order to minimise the misleading information occurring in the Conventional Wavelet.

2.6. The Introduction of the Reassignment Wavelet Transform

As mentioned before, Wavelet is a useful tool for analysing non-stationary signals. It can characterise signals over the time and frequency domain 0 0. However, it also contains a leakage problem in the energy representation. According to Tse and Yang 0 0, due to energy leakage and interference in the Conventional Wavelet, frequency overlapping and signal interference may occur. To improve the readability of the time-frequency analysis, the Reassignment method has been developed by Auger and Flandrin 0. The improved method contains the property of concentrating the detected signals in their respective frequencies and time locations. The resultant method will display the signals more precisely and become less affected by interference. This section will provide a brief discussion on Reassignment Wavelet. Then, Reassignment Wavelet will be used to analyse the machinery defects caused by impacts.

2.6.1. The Principal of Reassignment

The Reassignment technique contributes to the readability of energy distribution by moving every point on the time frequency plane to its centre of gravity 0. In other words, the Reassignment method gives an increased concentration of signal terms and at the same time cross terms can be effectively suppressed.

In Section 2.5.1, it was mentioned that the value of Wavelet Transform, at a given point, is a summarised value which is assigned to the geometric centre of the entire energy distribution. However, this energy distribution is not always geometrically symmetric. Figure 10 shows the time–frequency distribution of energy of an impulsive signal, and the rectangle is the resolution determined by Wavelet, whose geometrical centre is t_i, ω_i. It represents the averaged energy distributed in the window of resolution. However, there is actually no energy at this centre. Therefore, this averaging will cause interference and provide a lower concentration of the results.

Reassignment is a technique that changes this averaged geometrical centre to the centre of gravity, $(\hat{t}_i, \hat{\omega}_i)$, among the energy distribution. It overcomes the shortcoming of energy leakage and increases the concentration in the time-frequency plane. That is the spirit of Reassignment.

2.6.2. The Definition of Reassignment Wavelet Transform

The Reassignment Wavelet, $\emptyset_X$, transform is defined by

$$= \iint (\hat{a}/a)^2 |\, W_x(a,b;\, \psi)|^2\, \delta(\hat{b} - b'(ab))\, \delta(\hat{a} - a'(a,b))\, \delta a\, \delta b, \qquad (4)$$

where

$$b'(a,b) = b - \text{Re}\, \{a\, \frac{W_x(a,b;\, \psi\, \prime)\, W_x^*(a,b;\, \psi)}{|W_x(a,b;\, \psi)|^2}\}\ ;$$

$$\frac{\omega_0}{a'(a,b)} = \frac{\omega_0}{a} + \text{Im}\, \{\frac{W_x(a,b;\, \psi\, \prime)\, W_x^*(a,b;\, \psi)}{|W_x(a,b;\, \psi)|^2}\}\ ;$$

ψ is the function of a selected mother Wavelet, $W_x(a, b; \psi)$ is the CWT for a signal(x), a and b is the scale parameter and translation parameter in CWT respectively.

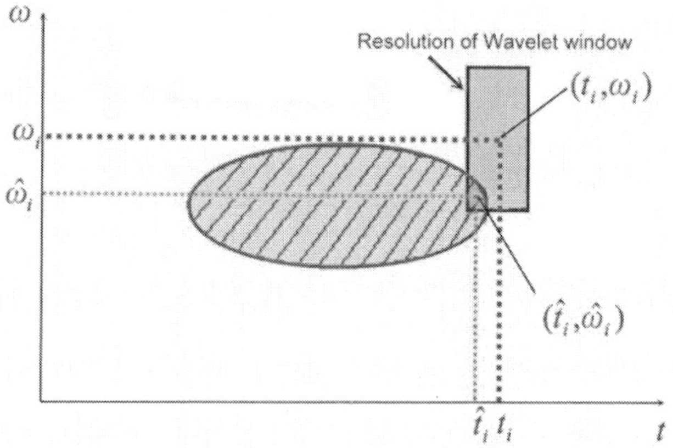

Figure 10. The principal of the Reassignment method 0.

2.7. The Use of the Higher Order Statistics in Machinery Fault Detection

In our daily life, there are many statistical indictors, such as the mean, variance and peak to peak value, which are very useful in machine fault diagnosis. In the previous section, the transducer and the methodology on vibration measurement were discussed. In this section, some of the useful waveform statistical analysis in vibration signals will be discussed. Furthermore, its theory will be stated and how it can be applied in machine fault detection will be examined.

2.7.1. Kurtosis

One mathematical representation of the deviation of an amplitude distribution from Gaussian is the so-called "fourth moment", or Kurtosis. Kurtosis is a statistic indicator, used in the time domain. According to Lorenzo and Calabro 0, Kurtosis is a good and reliable indicator of the real bearing running. One possible advantage of using Kurtosis as a fault detection parameter is that you do not need to have a trend over time for it to be effective. In most cases, when a faulty signal occurs, the Kurtosis value will go up. But the detected Kurtosis value will be wrongly enlarged due to the occurrence of random small impacts caused by the noisy signal. Therefore, it should be used together with Spectral RMS to avoid this kind of false alarm. The definition of Kurtosis is illustrated in Equation (5).

Definition of Kurtosis:

For the data $x_1, x_2 ... x_n$, the formula for Kurtosis is:

$$\text{Kurtosis} = \frac{1}{n}\sum_{i=1}^{n}\left(\frac{x(i)-\mu}{\sigma}\right)^4 \tag{5}$$

Where μ represents the average
σ represents the standard deviation
n represents the number of sample

According to Peled 0, Kurtosis is traditionally used as an indicator of damage in the area of bearing diagnostics. It represents three Gaussian signals, and deviates from zero for non-Gaussian ones, see Figure 11.

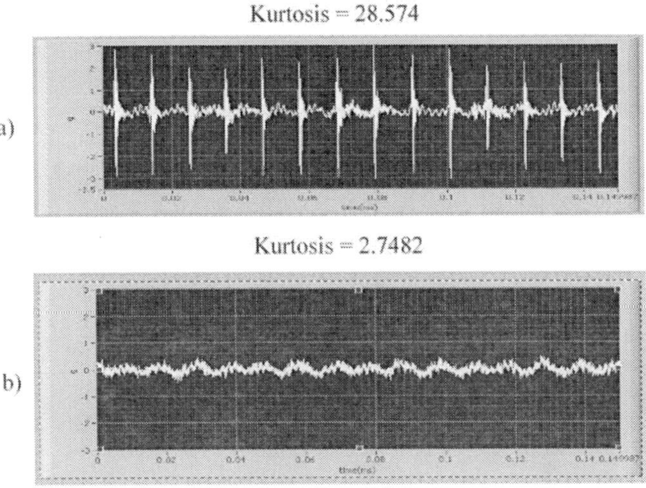

Figure 11. (a) A typical vibration signal generated by a faulty rolling element bearing (outer race fault), note the high Kurtosis value. (b) The same vibration generated by a normal bearing, the Kurtosis is relatively low.

2.7.2. Spectral Kurtosis

The Spectral Kurtosis is a fourth-order Spectral analysis tool recently introduced for detecting and characterising the impact in a signal. The definition of the Spectral Kurtosis was first introduced by Antoni and Randall 0, who generated the Kurtosis against frequency to wider classes of non stationary signals, as illustrated in Figure 12. The purpose of this section is to briefly review these techniques and motives, which led to the detection and characterisation of the impact that is caused by machinery faults.

Spectral Kurtosis is an advanced technique that allows the determination of the optimum frequency band and demodulates without historical data. According to Dywer 0, Spectral Kurtosis was originally devised to overcome the inefficiency of the power Spectral density, which is used to detect and to characterise transients in a signal. The idea is normally to calculate the Kurtosis at "each frequency band". In short, the Spectrum Kurtosis method first obtains a time-frequency diagram and then calculates the Kurtosis in the time direction of each frequency.

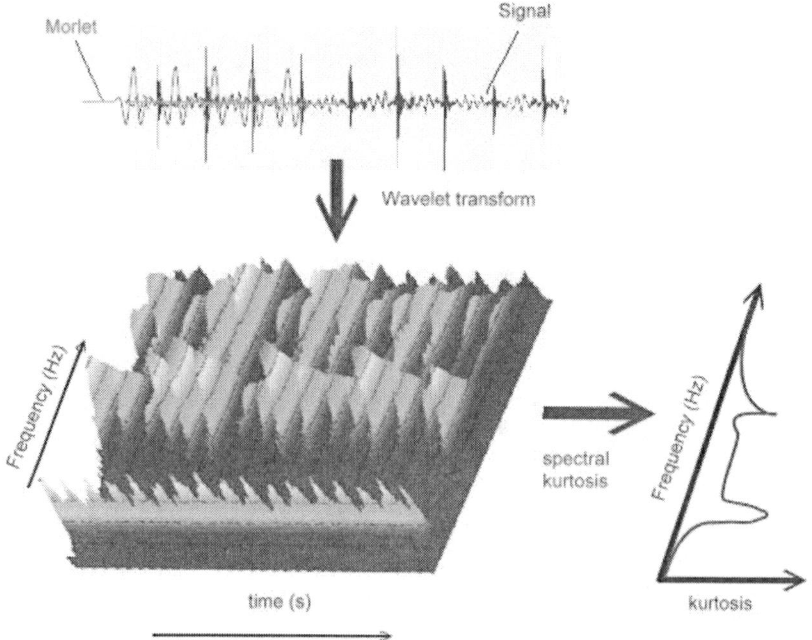

Figure 12. The generation of Spectral Kurtosis employs a Reassignment Wavelet Transform.

2.7.3. Root Mean Square

One of the most common vibration indicators is called the RMS value. According to Antoni and Randall 0, RMS can characterise the intensity of the signal. It is also the most relevant and simple measurement of amplitude. This is because it not only counts the time history of the wave, but also gives an amplitude value which is directly related to the energy content. Therefore, the RMS is commonly used as a destructive indicator. In terms of machinery, the RMS can be used as the fault indictor of the machine.

2.7.4. The Rationale of Proposing a Novel Spectral RMS x Kurtosis for Effective Bearing Fault Detection

When a bearing is running in normal conditions, the detected RMS values and Kurtosis values will be small because the overall vibration energy is small and fewer impacts appear respectively. In other words, when the bearing is becoming faulty and deteriorating, its overall vibration will increase. Lots of impacts will appear because the defective surface is continuously in contact with other surfaces when the bearing is rotating. Each contact will create an impact. Hence, the RMS and Kurtosis values are large in the case of a faulty bearing.

When a bearing signal is influenced by small noise, the overall vibration energy is small. Hence, the RMS value should be small. However, the detected Kurtosis value will be large due to the occurrence of random small impacts caused by the noisy signal. On the other hand, when a bearing signal is immersed in a very noisy environment, the overall vibration energy will increase, and eventually, the RMS value should be large. However, due to the large overall energy, the randomly occurring impacts will be either averaged or overwhelmed by this large white noise. Therefore, the Kurtosis value becomes small.

In conclusion, when a true defect exists in a bearing and is largely influenced by noise, the RMS x Kurtosis value should be distinctively higher than that occurs in a normal bearing. Therefore, a new faulty bearing detection method named Spectral RMS x Kurtosis has been proposed in this research. With the help of this novel method, the bearing excitation frequency zone, usually used for revealing the existence of impacts caused by a defective bearing, can be easily found. Moreover, based on the experience gained from diagnosing a bearing operating in a particular kind of condition, a threshold can be made so that the excitation frequency zone can be automatically found. Hence, the work load and the requirement of having an expensive expert for bearing fault detection can be eased and avoided.

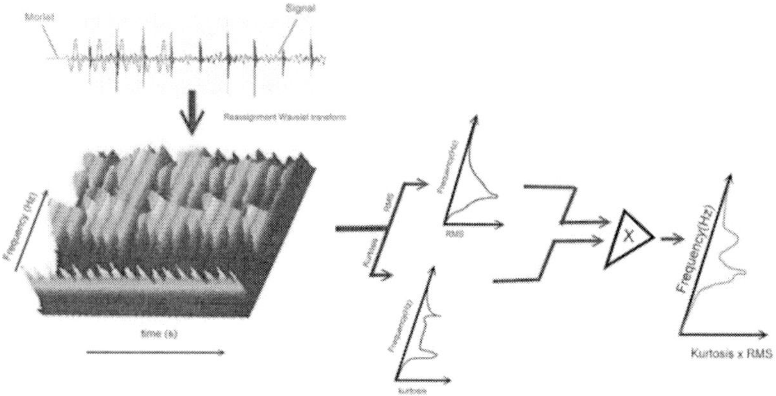

Figure 13. The generation of Spectral RMS x Kurtosis employs the Reassignment Wavelet Transform.

2.7.5. Spectral RMS x Kurtosis

As mentioned previously, RMS can provide a measurement of intensity, while the Kurtosis can indicate the impulsiveness of a signal. To combine the benefits of RMS and Kurtosis, a novel method named 'Spectral RMS times Kurtosis' or 'Spectral RMS x Kurtosis' has been designed and implemented. This is a new idea that calculates the product of the RMS value and the Kurtosis value at each frequency band, in order to reveal the impulsive signal and indicate its frequency band, that has been overwhelmed by other signals or frequency components generated by larger structures. The method is graphically illustrated in Figure 13.

Once the signal is collected, the Reassignment Wavelet can be used to generate the time and frequency spectrum. In this research, Morlet is employed and used as the mother function of Reassignment Wavelet. The Spectral RMS x Kurtosis will be treated by using this time-frequency spectrum after the time-frequency spectrum is generated. Through this time-frequency spectrum, the corresponding RMS values and Kurtosis values of each frequency band can be derived. As a result, two frequency spectra can be generated. One is the spectrum of RMS and the other is the spectrum of Kurtosis. Finally, the Spectral RMS x Kurtosis is obtained by calculating the product of the RMS value and the Kurtosis value of each frequency band.

2.8. The Evolution of Reassignment Wavelet Based Spectrum RMS X Kurtosis

In the past, FFT was used to detect machinery faults in the frequency domain. However, FFT does not function well for non-stationary signal 0, 0. Hence, Wavelet was developed. It divided the signal into different frequency sub-bands and provided a time representation for individual analysis, while FFT concentrated on the frequency magnitude only. Therefore, Wavelet is advantageous over traditional FFT methods in analysing vibration signals. Meanwhile, Wavelet suffered from energy leakage 0. This energy leakage interrupts the identification of the bearing excitation zone and causes a failure of fault detection.

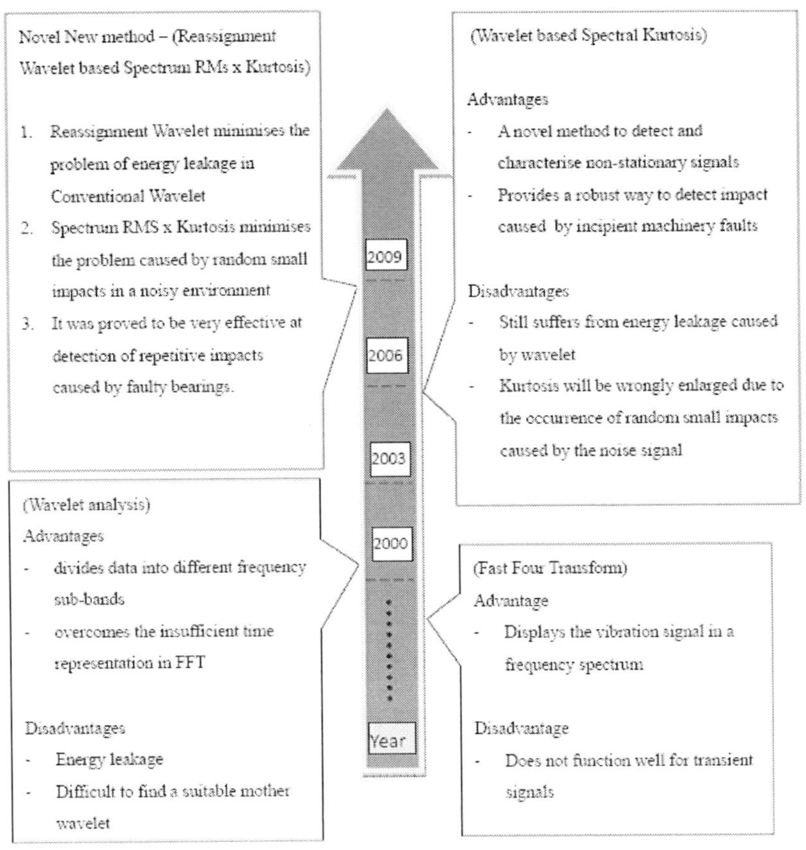

Figure 14. The evolution of the proposed new method.

Later, Spectral Kurtosis was developed, which is a novel method to detect and characterise the impact signal caused by incipient machinery faults 0, 0. However, at that time, Spectral Kurtosis still suffered from the leakage problem due to the application of Wavelet. Also, Kurtosis was not appropriate to indicate the machinery fault when it was influenced by a strong background noise 0.

Finally, a new method, namely Reassignment Wavelet based Spectral RMS x Kurtosis, has been designed. By adopting the method of Spectrum RMS x Kurtosis, the problem of noise influence to Kurtosis has been minimised. Also, the leakage problem existing in the Conventional Wavelet has been improved by using the Reassignment Wavelet because the Reassignment method contributes to the readability of the energy distribution by moving every point on the time frequency plane to its centre of gravity. The evolution of this proposed new method is illustrated in Figure 14.

Section 3. The Design of the Virtual Based Automatic Fault Detection System

As the number of monitoring points and complexity of fault detection increases, a computer-based virtual instrument will be the most economic solution. A collection of vibration samples from each machine can be organised by the computer. A complementary computer-based system with a virtual software instrument was developed in this research. A major benefit of this system is that the entire vibration signal is permanently preserved in the hard drive. Also, it makes it easy to perform future enhancements or modifications, due to its completely virtual based developing platform. In the following chapter, the implemented virtual instrument will be discussed as well as its data acquisition and analysing functions.

3.1. Development of a Single Tasked Data Acquisition Program

To aid the test, a simple data acquisition program has been designed. LABVIEW was chosen as the development platform. It provides a graphical development environment for the programmer. By using the graphical platform, the methodology can be created and modified rapidly. Regardless of experience, LABVIEW makes development fast and easy to do.

Figure 15, shows the data acquisition panel of the designed system. The input signals are sampled in accordance with the prescribed sampling rate, number of samples, sensor sensitivity and their gain. Once the prescribed data have been acquired, the sampled temporal waveforms will be displayed in one of the four channels. The data collection process can be continuous and all four waveforms will be updated momentarily.

3.2. Implementation of the Reassignment Wavelet Analysis

The Reassignment Wavelet analysis system equips Morlet as the mother Wavelet. These algorithms help to identify the defect type as the Reassignment Wavelet graph points out the time, magnitude and frequency of the impact signal. By using the Reassignment Wavelet, the component defects and the types of defects, such as an outer race defect, inner race defect and ball defect, can be identified easily.

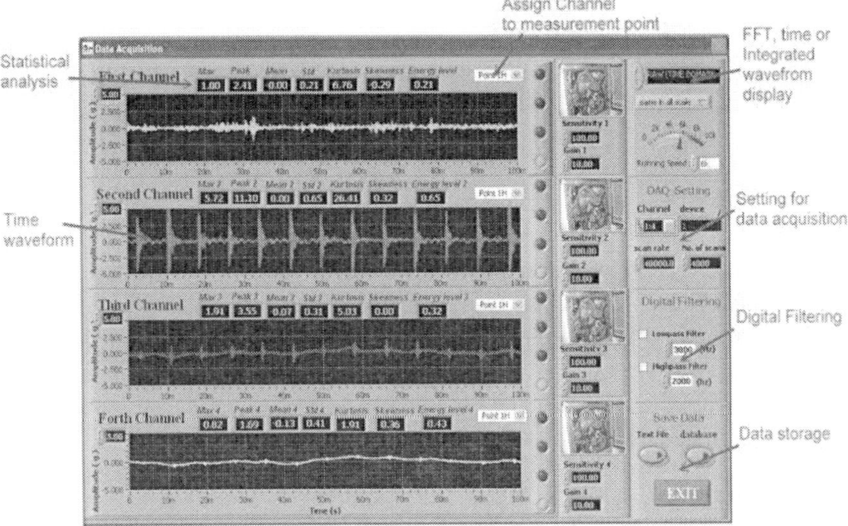

Figure 15. The designed fundamental DAQ and signal processing panel.

The designed Reassignment Wavelet panel is shown in Figure 16. The figure consists of the time waveform along with its FFT and Reassignment Wavelet transform plots. From here, two list boxes, namely the intensity graph and the multi-layer display can be selected. It provides an overall picture of the machine's health. The latter helps to reveal faults in different machine components.

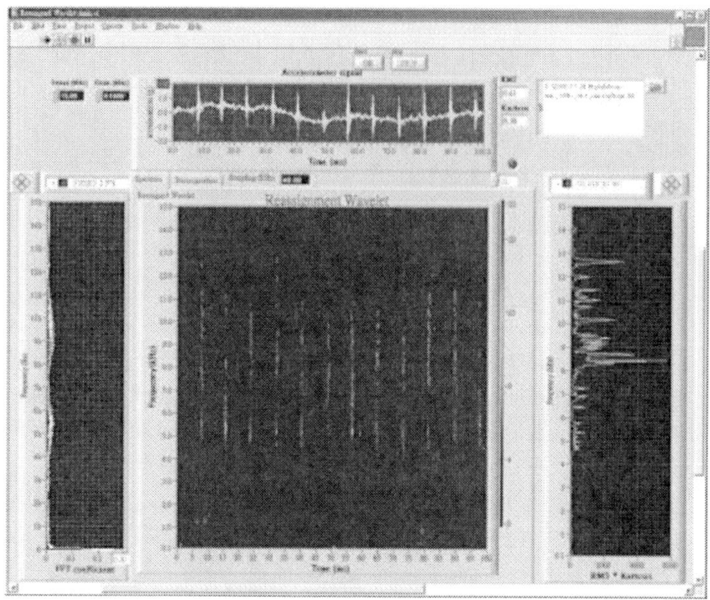

Figure 16. The designed virtual instrument for performing Reassignment Wavelet analysis and Spectral RMS x Kurtosis.

Once the calculation is processed, a green LED illuminates until the analysis is complete. As soon as the analysis is complete, the Reassignment Wavelet Transform plot and the Spectral RMS x Kurtosis are displayed. In the Reassignment Wavelet plot, the impact strength is represented by colours. The highest and lowest scales are indicated in red and blue, respectively.

3.3. Screen Flow Design and Functionality

In the previous section, the literature relating to machine fault diagnosis and its methodology was discussed. At this time, a sufficient virtual instrument was developed, which included the data acquisition, data storage and analysis function. Figure 17 shows the front panel of the designed system. It includes a number of virtual instruments for the use of diagnosing different types of fault. The design of these virtual instruments was supervised by Dr. Peter Tse and was the collective work of his research team at the SEAM laboratory. These virtual instruments have been extensively used for monitoring the health of various industrial machines.

3.4. Time Domain Analysis

The time series signal can be used to perform fault and failure diagnosis by analysis of the vibration data obtained from the equipment. Some statistical values, such as RMS, peak, mean, standard deviation and Kurtosis, are widely used to investigate the random characteristics of the physical system as shown in Figure 18. These can provide useful information for detecting bearing faults and are commonly used in machine health monitoring. Hence, those statistical indicators are embedded in to the designed system.

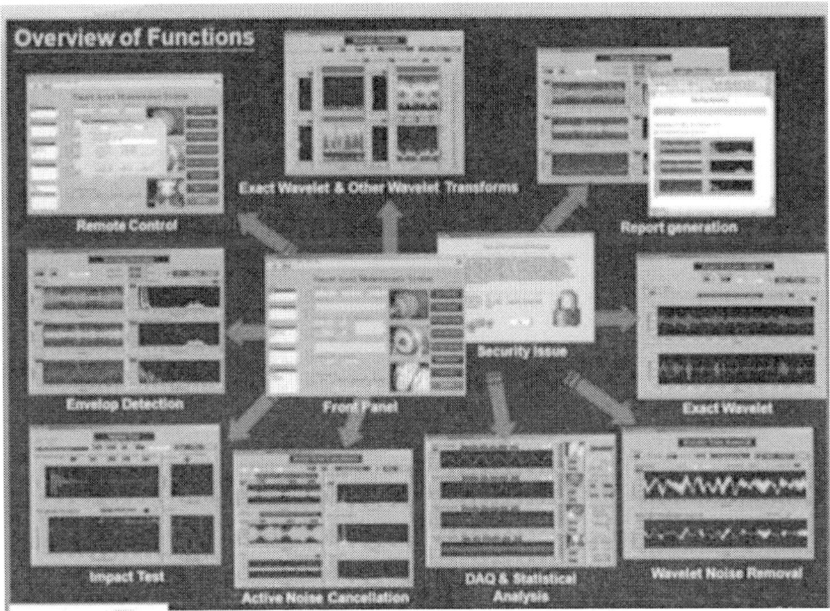

Figure 17. The designed intensive fault diagnosis system.

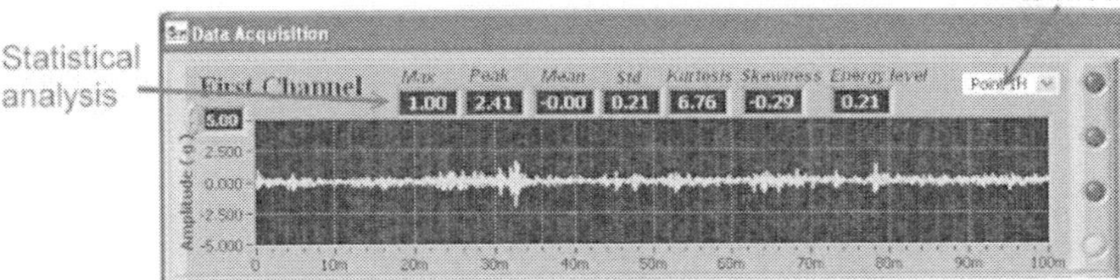

Figure 18. Selected statistical indicators under time domain analysis.

3.5. Data Storage and Extraction

The database is one of the major components for performing condition monitoring. Without a database, data would be difficult to organise or extract. Also, a well-organised data structure should include information about the sampling date, sensor position, measuring unit and characteristics of the machinery being measured. In this research, the Microsoft Access based data recording system was implemented in order to record such kinds of information efficiently.

The database platform was designed under the supervision of Dr Peter Tse and his SEAM research team. It provides a graphical user interface (GUI) allowing the researcher to choose from different diagnostic techniques, for example: Reassignment Wavelet Transform and Spectral RMS x Kurtosis. In addition, this system contains a comprehensive set of tools for acquiring, analysing, displaying, publishing and reporting, as well as a database for the storage and retrieval of the acquired machine data. Figure 19 shows a typical GUI data record. On this panel, tick boxes, digital indicators, control boxes, LED indicators, push buttons, and switches were integrated to peruse the basic requirements for performing machine fault diagnosis.

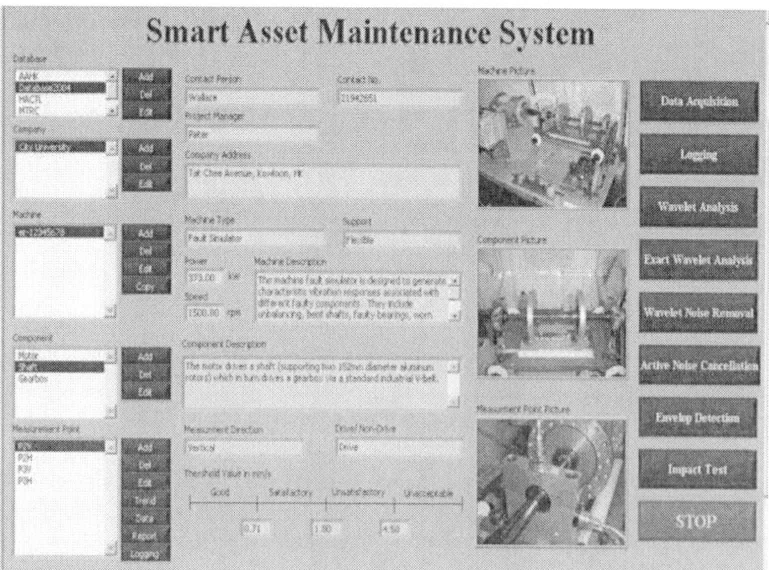

Figure 19. The database panel in the system.

The database enables the researcher to analyse the vibration signal. The hierarchy of the designed database consists of 5 entities, namely, the company identification, the machine identification, the inspecting component, the measurement location orientation, and the original data file. In addition, a visual data explorer and data manager is included in this virtual instrument, as illustrated in Figure 20.

After the creation of the database, another effective tool in this system is the Data Manager. By using the function of the Data Manager, all the vibration data will be listed according to their sampling dates and times. The data history can be revealed for further investigations of machinery health. The database tools would contribute to data collection and extraction during the experiment.

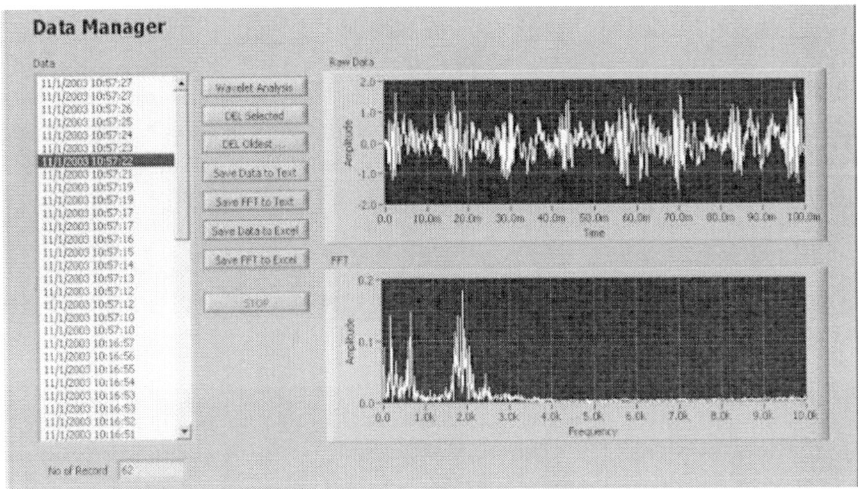

Figure 20. The designed data management tools.

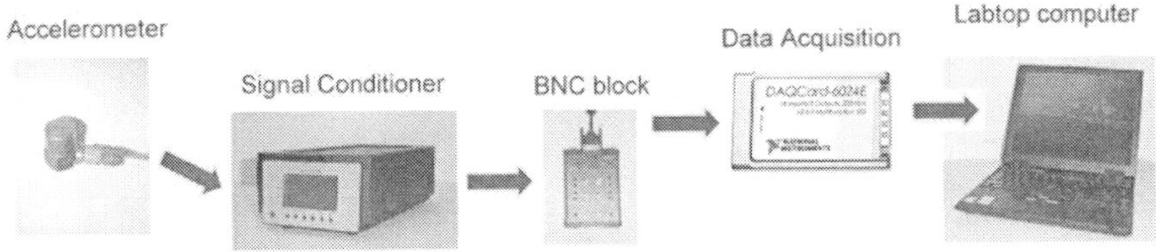

Figure 21. The layout of the set up.

3.6. The Layout of the Hardware Configuration

To acquire the vibration signals generated from an inspected machine, a number of devices were employed. Figure 21 shows all the devices that were used with the virtual designed instrument. The signals collected by the accelerometers were conditioned by a multi-channel signal conditioner prior to the input of the data acquisition (DAQ) card. The DAQ card was used because it converted the conditioned vibration signals into a digitized format. Also, the DAQ card acted as a communication interface between the laptop computer and each physical channel. Finally, these conditioned digital signals were stored on a hard disk via the computer. In short, the computer became a physical analyser for fault diagnosis.

The proposed system has been implemented based on the configuration as shown in Figure 21 by using a multiplexing data acquisition card as well as an interfacing board. The maximum sampling frequency of the data acquisition device is 250k Hz, which means that 250,000 units of data can be sampled every second. On the other hand, the interfacing board provides multi channels of analog to digital converters for accessing different types of accelerometer.

Section 4. Experiment on Laboratory Machinery Fault Simulator

This chapter provides the experimental results for bearing fault detection. It contains three sections. The first section is about the bearing fault detectioin. In the laboratory, the Reassignment Wavelet is used to analyse experimental data for three typical faults of rolling element bearings: ball defects, outer race defects and inner race defects. The second section is about the extraction of a faulty bearing signal from a noise submerged environment.

The adopted apparatus for this project is discussed below. The setup of the bearing fault simulator, the construction of the rolling element bearing and the principle for calculating the bearing characteristic frequency are illustrated in this section.

4.1. The Bearing Fault Demonstrator

The test rig used in the laboratory consisted of a bearing fault demonstrator. It was designed to simulate different types of bearing defects, such as an outer race defect, inner race defect and ball defect. As shown in Figure 22, the demonstrator incorporates one rolling element bearing, one transducer and one single phase motor. The inspection bearing was installed into the bearing house. It directly connected with the bearing without using any coupler or external joint. Hence, the experiment assumed that there would be no misalignment between the motor and the bearing. Bearings with normal and damaged conditions were demonstrated separately. The transducers for measurement of vibration were mounted vertically with respect to the rotating shaft. During this research, a new diagnostic system was implemented, which was used to analyse the signal collected from both the laboratory and industrial machines.

The laboratory test, is a control experiment, and should be done in an ideal environment. The testing equipment used in the laboratory was put on a bench. When the motor started running, the bench vibrated as well. Therefore, in this test, a rubber sheet was used to isolate the test equipment to reduce the vibration transmission from the table to the testing bearings. In this way, the laboratory experiment concentrated only on the vibration emitted from the faulty bearing. For the industrial test, industrial partners did not provide isolation for testing purposes. But since the machine size is much larger than the equipment used in the laboratory, and they are positioned on flat ground instead of a laboratory bench, the problem is minimised.

The schematic diagram of the data collection is shown in Figure 23. The transducer used was directly powered by a signal conditioner. The conditioned vibration signal was collected and stored in a separate computer via a data acquisition card. At the end of this work, some results on bearing faults will be discussed, which were compared based on the same circumstances.

Figure 22. A bearing fault simulator.

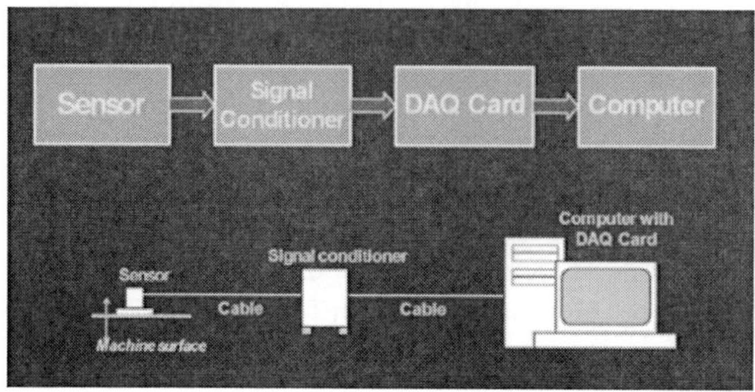

Figure 23. A schematic diagram for measuring the bearing fault using a fault simulator.

4.2. The Rolling Element Bearings

Rolling element bearings are one of the most widely used elements in machines. They are easy to find in every kind of machinery and they transmit forces between rotating machine components. Their presence can be found in lots of applications, such as motors, gas turbines and pumps. Bearings have received a lot of attention in the field of vibration analysis as they represent a huge source for early fault detection. Consequently, bearing failure is one of the most common causes of breakdowns in most rotating machines. Failure could be initiated as a result of manufacturing defects such as micro defects on the rolling element, or as a result of improper installation practices.

The damage in typical bearing faults often consists of one of three things: damage to the outer-race surface, the inner-race surface and the surfaces of the rolling element, as illustrated in Figure 24. This section will concentrate on the methods used to determine the types of bearing defects.

Table 1. presents three typical examples of rolling element bearings used in this research. Although only three types of bearing are mentioned in this thesis, the proposed new methodology was tested on many other bearings, e.g. roller bearings and general bearings, the results were positive, too. The proposed method can also be applied to different kinds of mechanical components, such as pumps or gears. However, due to limited time and resources, this thesis has concentrated on examining bearings only.

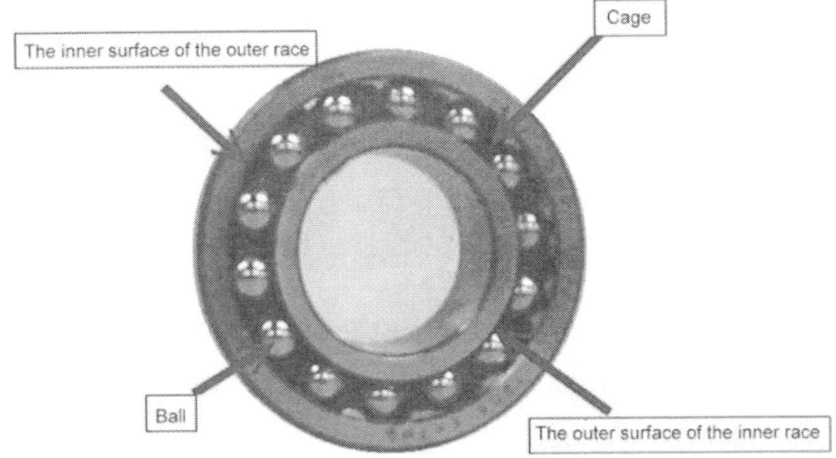

Figure 24. The rolling bearing structure.

Table 1. The three examples of bearings examined in this research

	Laboratory Fault Simulator	**Railway Traction Motor**	**Cooling Water Pump**
Model	SKF 1206	SKF 6215	Koyo 62052RS
Type	Deep groove ball bearing	Deep groove ball bearing	Deep groove ball bearing
Ball diameter	8 mm	17.6 mm	9 mm
Pitch diameter	46.55 mm	104.55 mm	41 mm
No. of rolling elements	14	11	10
Loading	No external loading	250 kg rotor	No lateral load

Table 2. The equation for calculating the bearing characteristic frequency

Train (cage) frequency (TF):	$\frac{1}{2}f_r\left(1-\frac{d}{D}\cos\beta\right)Hz$
Ball pass frequency outer (BPFO):	$\frac{n}{2}f_r\left(1-\frac{d}{D}\cos\beta\right)Hz$
Ball pass frequency inner (BPFI):	$\frac{n}{2}f_r\left(1+\frac{d}{D}\cos\beta\right)Hz$
Ball spins frequency (BSF):	$\frac{D}{d}f_r\left[1-\left(\frac{d}{D}\right)^2\cos^2\beta\right]Hz$
	β = angle of contact (β=0 for ball bearing) n = number of rolling elements f_r = relative rotation frequency d = rolling element diameter D = pitch diameter

The theoretical foundation of bearing failure modes has been developed over the years by many researchers 0, 0 and 0. According to Ohta 0, it is necessary to assume ball bearing motion to be simple and only concentrate on their basic motions, in order to determine the characteristic frequencies of bearings. Table 2 describes the calculation of bearing characteristic frequency, which includes the Train (cage) frequency (TF), Ball Pass Frequency Outer (BPFO), Ball Pass Frequency Inner (BPFI) and Ball Spins Frequency (BSF).

By referring to the equations described in Table 2, it can be seen that theoretically the dimension of a bearing does not affect the capability of the proposed new method (Reassignment Wavelet based Spectral RMS x Kurtosis). The majority of the Reassignment Wavelet based Spectral RMS x Kurtosis is the retreading capability of the repetitive impact signal that is generated by the faulty bearing. As mentioned before, although only 3 types of bearing were described in this thesis, many other bearings were tested during the study period of this research. The results generated by these bearings were positive, too.

4.3. Artificially Induced Bearing Defects

To demonstrate the fault generated by a faulty bearing, some defective bearings were made artificially: they included bearings with outer race defects, inner race defects and ball defects. They are shown in Figure 25, Figure 26 and Figure 27 respectively.

Figure 25. Etched artificial crack on bearing outer race.

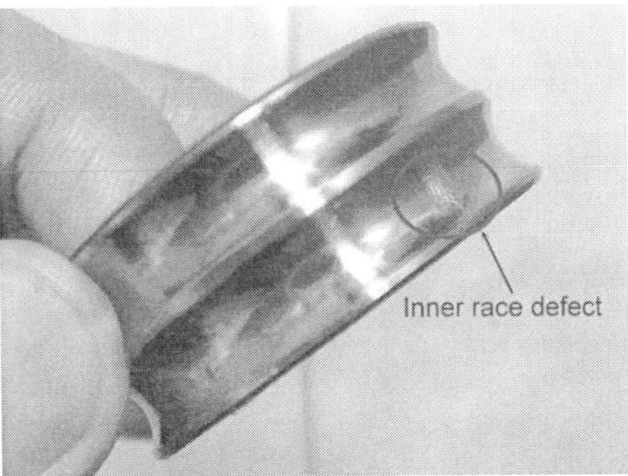

Figure 26. Etched artificial crack on bearing inner race.

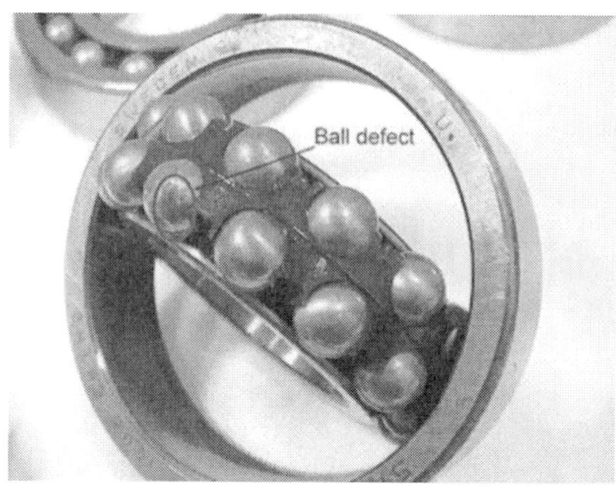

Figure 27. Etched artificial crack on rolling element.

4.4. A Comparison Study of Conventional Wavelet and Reassignment Wavelet

With the help of a virtual instrument, a Morlet based Wavelet spectrum was generated for analysing localised impact with a time and frequency series. The energy density is represented by different shades of colour. Compared with FFT, the Wavelet shows all components clearly, and the approximate time for the appearance of an impact is also shown. But, the excitation frequency caused by the defective bearing is still not clear. This is because the spectrums of Wavelet overlapped each other and the interference terms appeared on the time-frequency plane. Therefore, the results generated by Conventional Wavelet and Reassignment Wavelet were compared to verify the effectiveness of Reassignment Wavelet. The comparison results, as illustrated in Table 3, indicate that the Reassignment Wavelet can reflect the inherent features of the impacts caused by the defective bearing and can provide much clearer information for bearing fault detection. The comparison studies include a normal bearing and three commonly occurring defects in similar kinds of bearings. The results are illustrated as colour spectra generated by a bearing having an inner race defect (BPFI), a bearing having a cracked ball (BSF), a bearing having an outer race defect (BPFO), and a normal bearing (Normal). For reference purposes, the spectrum generated by a motor running without a bearing (Motor running without bearing) is also included in the comparison study.

Table 3. A comparison between Conventional Wavelet and Reassignment Wavelet with different types of bearing defects and running conditions

Bearing defects	Conventional Wavelet	Reassignment Wavelet
BPFI	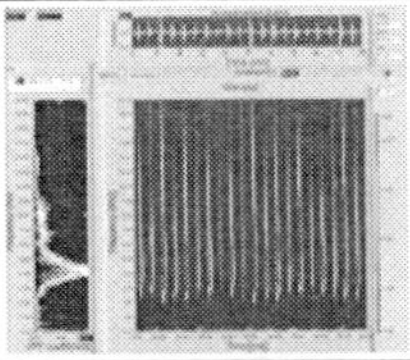	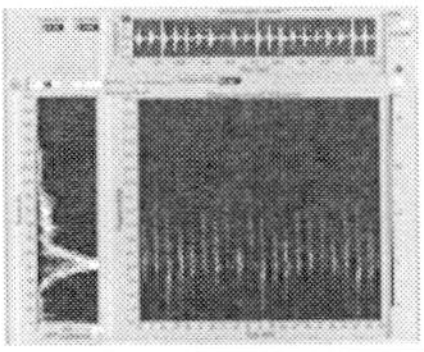
BSF	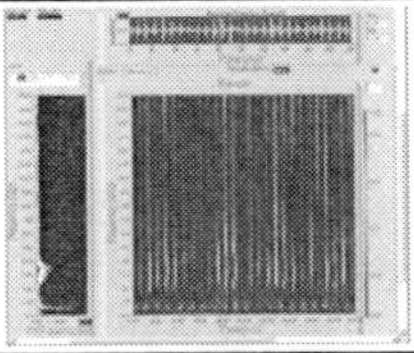	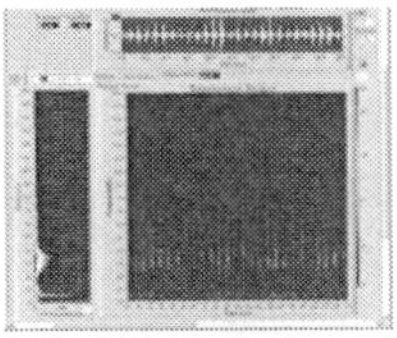
BPFO	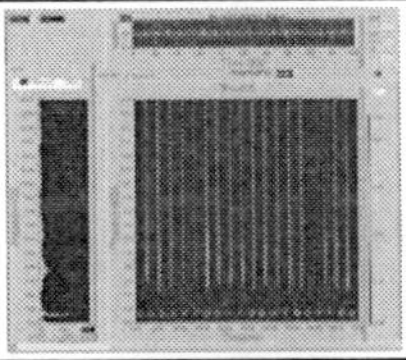	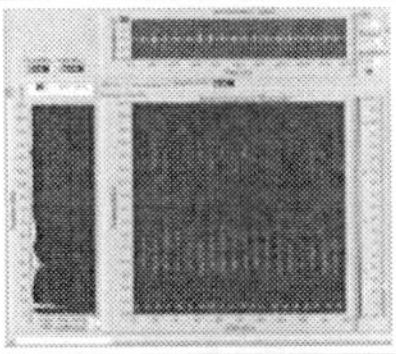
Normal	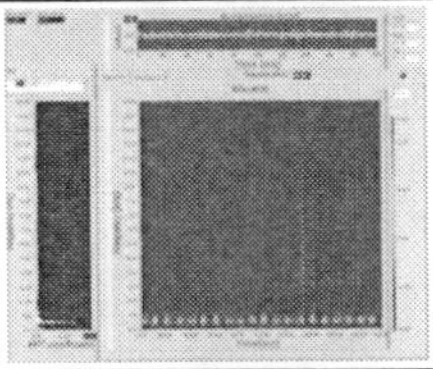	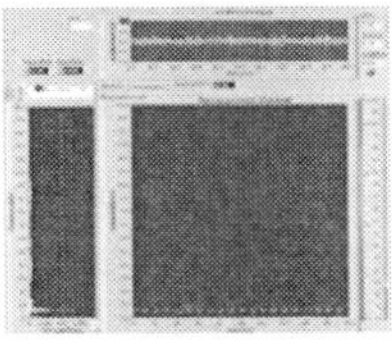
A motor running without the bearing	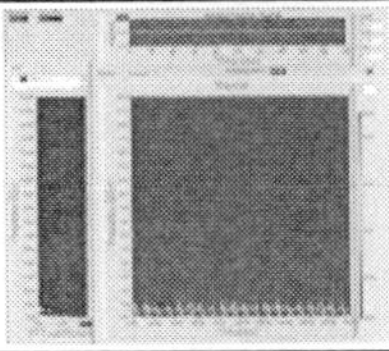	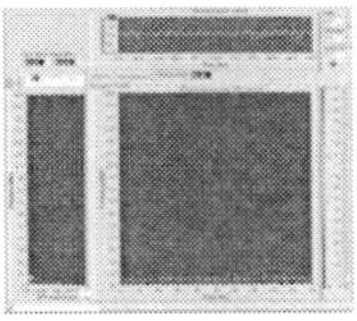

4.5. Bearing Fault Detection by Using RMS and Kurtosis

To verify the concepts proposed in Section 2.9.4, a series of experiments were conducted to investigate the effectiveness of RMS and Kurtosis in bearing fault detection. A set of experiment data were collected for each case when the bearing was running normally under small noise and large noise environments, and with a bearing suffering from an outer race defect.

Figure 28, shows a bearing signal running in normal conditions. The detected RMS value and Kurtosis value are 0.09 and 2.73 respectively. In this situation, the overall vibration energy is small and no impact occurs in the collected signal. Meanwhile, Figure 29 presents a temporal signal collected from a defective bearing. The detected RMS and Kurtosis values are 0.31 and 28.5 respectively. Both of the RMS and Kurtosis values are higher than those values obtained under normal conditions. This is because the overall vibration is high and lots of impacts occurred in the signal.

For the sake of comparison, the same normal bearing signal was submerged in a small noisy environment as shown in Figure 30. The detected RMS and Kurtosis values are 0.1 and 12.82 respectively. In this case, the detected RMS is small while the detected Kurtosis is large. This is because the overall energy is low and random impacts with small magnitudes exist. On the other hand, when the same normal bearing signal was running in a very noisy environment, as shown in Figure 31, a large value of RMS [0.93] appears because of the increase in the overall energy. However, the Kurtosis value [3.03] detected in this condition is small because the impacts have been overwhelmed by the high level of noise.

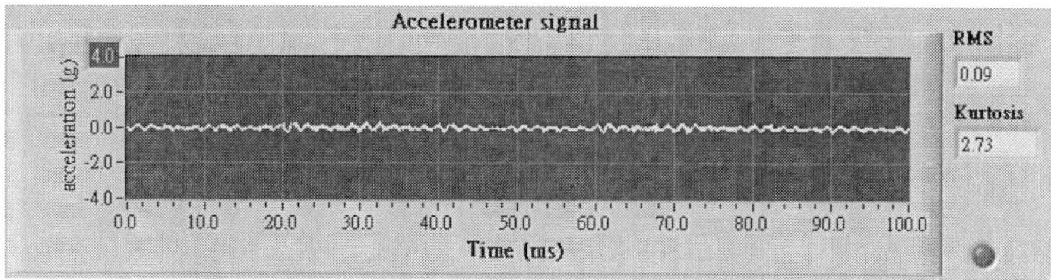

Figure 28. Bearing signal running in normal conditions.

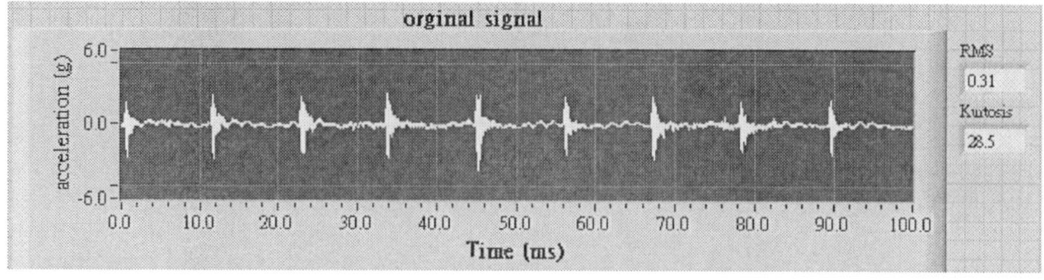

Figure 29. True defective bearing (outer race bearing).

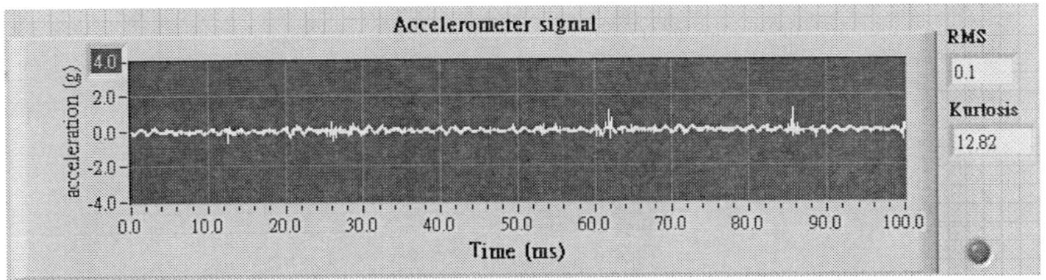

Figure 30. Normal bearing signal submerged in a small noise environment.

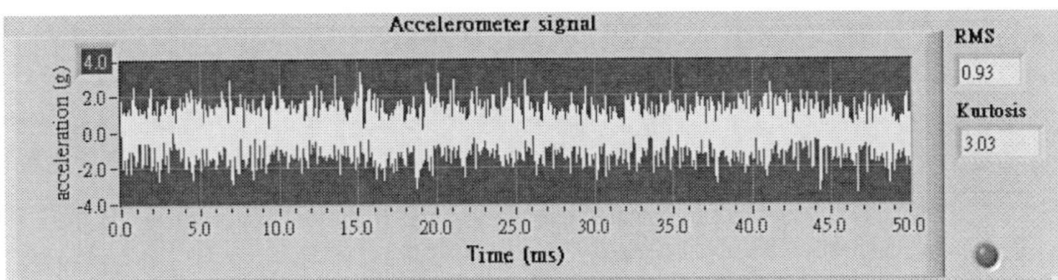

Figure 31. Normal bearing signal interference by a high level of noise.

According to the above experimental results, the effectiveness of RMS and Kurtosis in bearing fault determination, previously described in Section 2.9.4 of the literature review, has been verified. Encouraged by such results, the next development was to investigate the feasibility of detection of the bearing excitation zone by applying the Reassignment Wavelet based Spectral RMS x Kurtosis.

4.6. Detection of a Normal Condition Signal and Motor Signature

In this section, the Reassignment Wavelet, coupled with Spectrum RMS x Kurtosis, will be used to analyse the experimental data of three typical faults in rolling element bearings. In Figure 32, data was obtained under normal machine conditions. Provided that the bearing is new and in good condition, it serves as a baseline for bearing condition in future studies. Figure 33 shows data obtained under the condition that the bearing was removed from the rotating shaft. Only the motor and the bearing house were measured. The motor vibration signature was then detected for future reference.

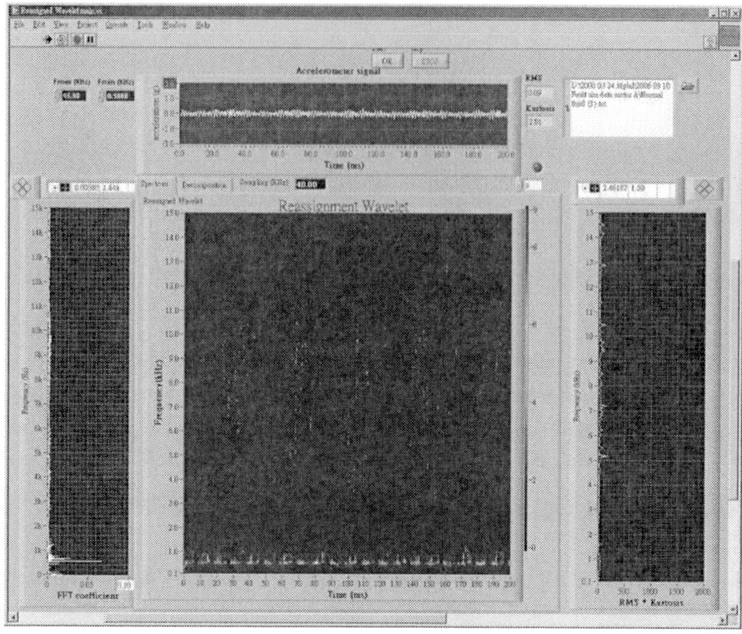

Figure 32. A reference signal under normal conditions.

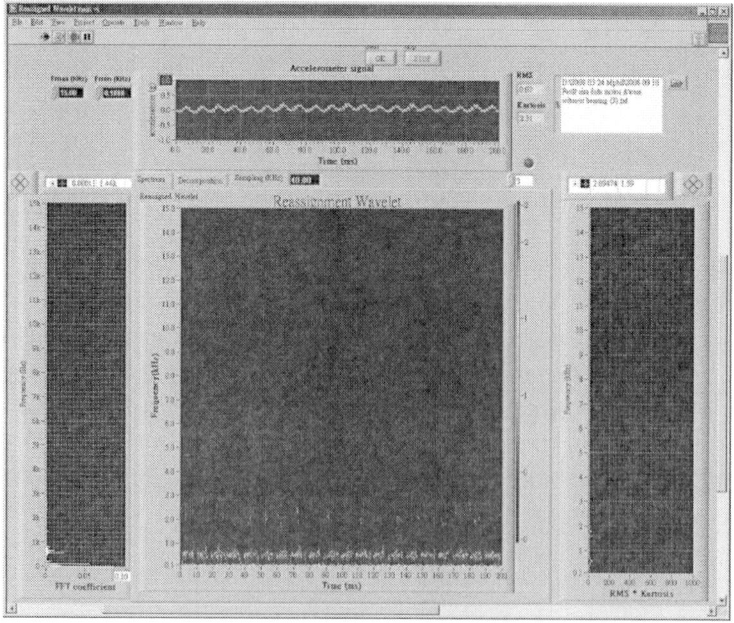

Figure 33. A motor vibration signature for future reference.

4.7. The Analysis of the Bearing Signal Collected from a Bearing with a Ball Defect

A defect was artificially made by inserting a cracked ball into the bearing. A set of vibration data was sampled at 40 kHz. The shaft rotational speed was set at 1400 rpm [23.3Hz]. According to Table 2, as previously illustrated in Section 3.3.2, the bearing defective frequencies can be calculated via the ball bearing dimensions, which are directly related to the shaft rotational speeds. These values are calculated and presented in Table 4.

The vibration temporal signal of a cracked ball bearing is shown in the top diagram of Figure 34. There are several periodic impacts which exist in the time waveform. The time interval was found to be 7.6 ms after zooming two adjacent impacts, as shown in the bottom diagram of Figure 34. The observed time interval is close to the calculated time interval for BSF, which is 7.65 ms as tabulated in Table 4.

Table 4. Calculated bearing characteristic frequency

	Calculated Frequency
Train (cage) frequency (FTF):	= 9.57 Hz (104.53 ms)
Ball pass frequency outer (BPFO):	= 135.33 Hz (7.39 ms)
Ball pass frequency inner (BPFI):	= 191.33 Hz (5.23ms)
Ball spins frequency (BSF)	= 130.66 Hz (7.65ms)

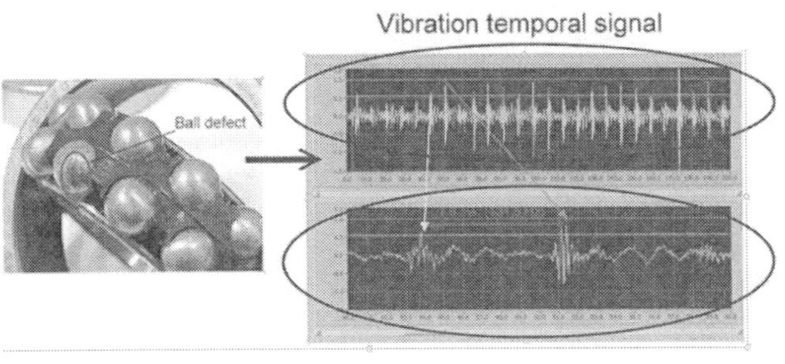

Figure 34. The temporal bearing signal collected from a bearing with a ball defect.

Because of the benefits of using Reassignment Wavelet mentioned in the previous section, the Reassignment Wavelet was adopted to generate the Spectral RMS x Kurtosis in further investigations. The result is presented in Figure 35. The left side of Figure 35 is the FFT spectrum of the defective bearing signal (ball defect), through which it can be seen that the energy mainly centralises at 545Hz, while other components are small and are not shown clearly in the FFT spectrum. Compared with the FFT spectrum, the Reassignment Wavelet (middle of Figure 35) shows all components clearly, and the approximate time for the appearance of an impact is also shown. Therefore, the concentration of signal components was enhanced.

The Reassignment Wavelet shows that the impacts exist between 3.5 kHz and 4.5 kHz. To further investigate the exact frequency band of the impacts, Spectral RMS x Kurtosis was used. The right hand side of Figure 36 shows the Spectrum RMS x Kurtosis of the ball defect bearing signal. It shows that the maximum RMS x Kurtosis value is 4.022 kHz. This result matched with the result calculated by the Reassignment Wavelet. In addition, six decomposed layers are selected and presented in the middle of Figure 36. Layer 5 [4022 Hz], highlighted with red circles, contains the highest value of RMS x Kurtosis and lots of impulsive signals can be detected in this layer.

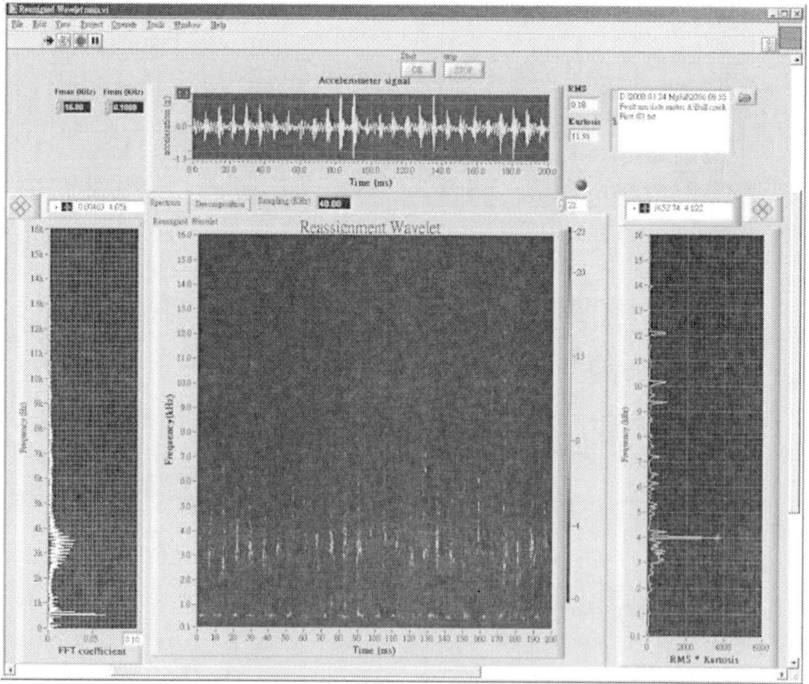

Figure 35. Reassignment Wavelet analysis on a ball defective bearing.

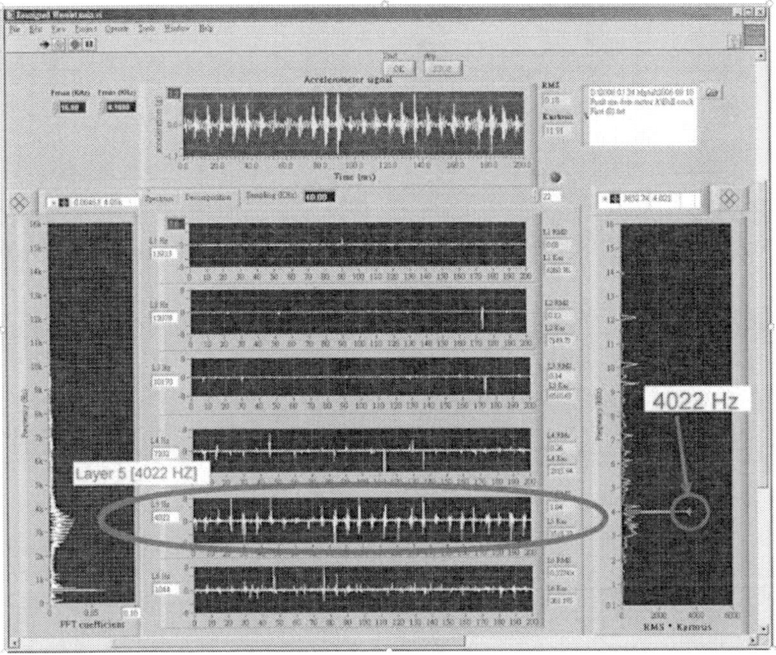

Figure 36. Decomposed layers of bearing signals collected from a bearing with a ball defect.

4.8. The Analysis of Bearing Signals Collected from Bearings with Outer Race and Inner Race Defects

In this section, the bearing signals collected from bearings with inner race and outer race defects were under investigation. Both defects were artificially made. Cracks were created on the inner race surface and outer race surface respectively. The impact signals caused by these inner race and outer race defects are shown in Figure 37 and Figure 38 respectively.

Again, the Reassignment Wavelet was adopted to generate the Spectral RMS x Kurtosis. The frequency layer of 4.123k Hz that contains the largest value of Spectrum RMS x Kurtosis was detected for the case of an inner race defect, while this was detected at 4.719 kHz for the case of an outer race defect. The detected frequencies and layers are presented and highlighted with red circles in Figure 37 and Figure 38. Thus it can be seen that the detected layer can show the impact explicitly.

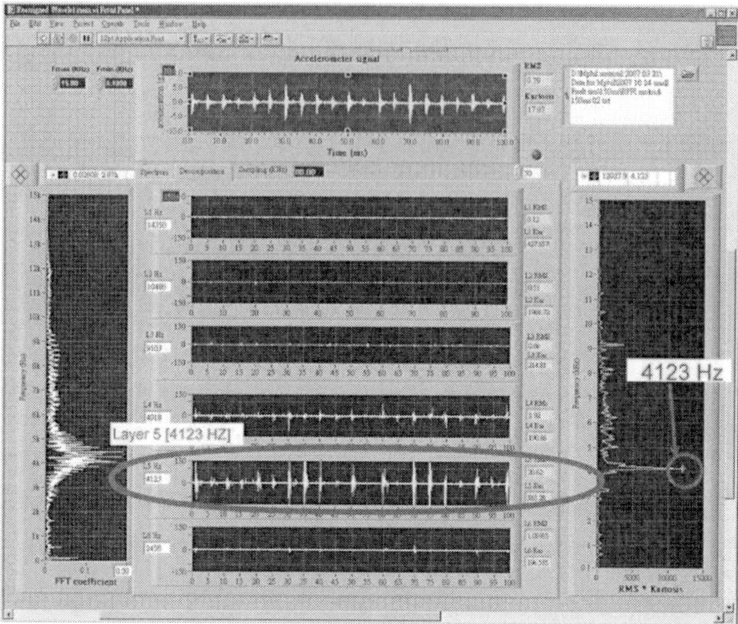

Figure 37. Decomposed layers of the bearing signal with inner race defect.

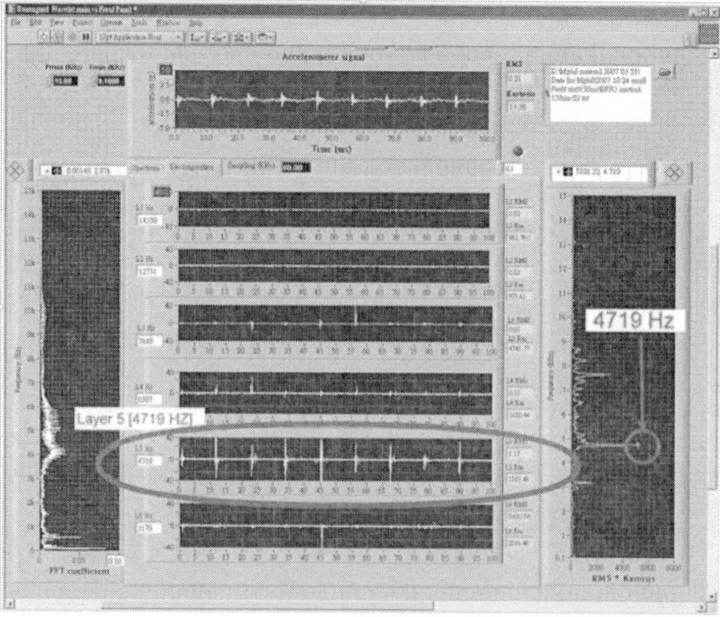

Figure 38. Decomposed layers of the bearing signal with an outer race defect.

The laboratory test is a controlled experiment to demonstrate the true bearing defects. In other words, the laboratory experiment was trying to obtain the "signature of the bearing defects" for future comparison. Figure 36, Figure 37 and Figure 38, present the results of using spectral RMS x Kurtosis for analysing three typical bearing faults. They include a ball defect, outer-race defect and inner-race defect. The detected frequencies and corresponding frequency layers, by using Spectral RMS x Kurtosis, are highlighted with red circles. Through this figure, the impact of the detected layer was shown explicitly. Therefore, Spectral RMS x Kurtosis worked in the laboratory experiment.

4.9. Experimental Analysis with Computer Generated Noise Simulation

A Gaussian white noise signal was generated by a virtual instrument. It was used to mask a well known outer race defective bearing vibration signal, as illustrated in Figure 39. The computer-generated Gaussian white noise is shown in Figure 40. Then, the detected outer race bearing defective signal is merged with this generated Gaussian white noise and is presented in Figure 41.

The Reassignment Wavelet was employed to analyse this merged signal. The result is presented in Figure 42. Figure 42 shows that the impacts can no longer be seen by using both the FFT and Reassignment Wavelet methods. However, by calculating the Spectral RMS x Kurtosis, the impacts can be detected again.

In Figure 43, six layers of frequency peak shown on the Spectral RMS x Kurtosis are selected for further analysis. Decomposition analysis was used to analyse these selected layers. Corresponding RMS and Kurtosis values are shown in Figure 43. Frequency layers were then ranked according to the RMS and Kurtosis values. The layer with the largest value was ranked as 1 while the layer with least value was ranked as 6. A summary of the results is given in Table 5.

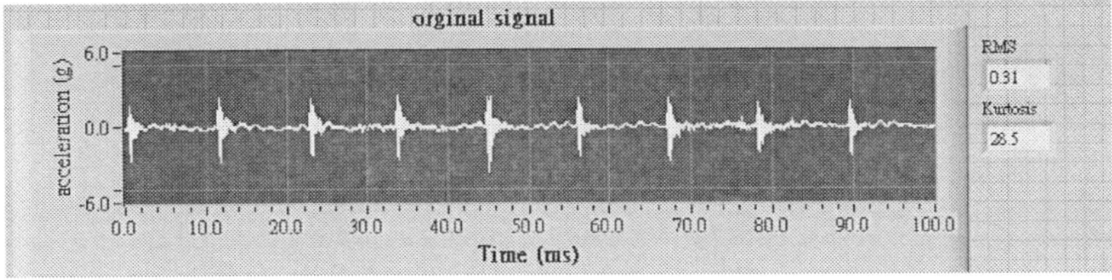

Figure 39. An outer race defective bearing signal.

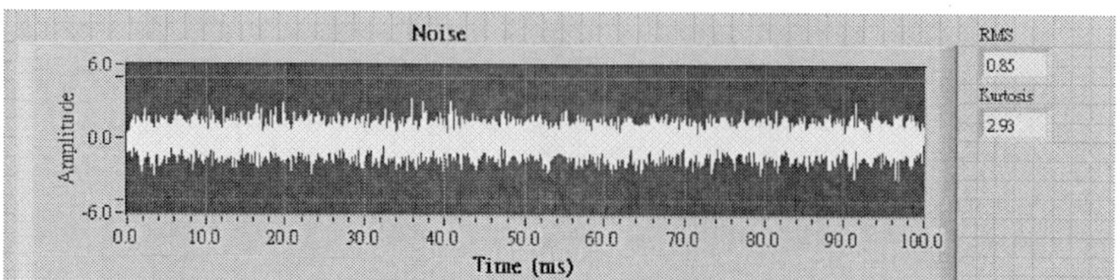

Figure 40. A computer generated Gaussian noise signal.

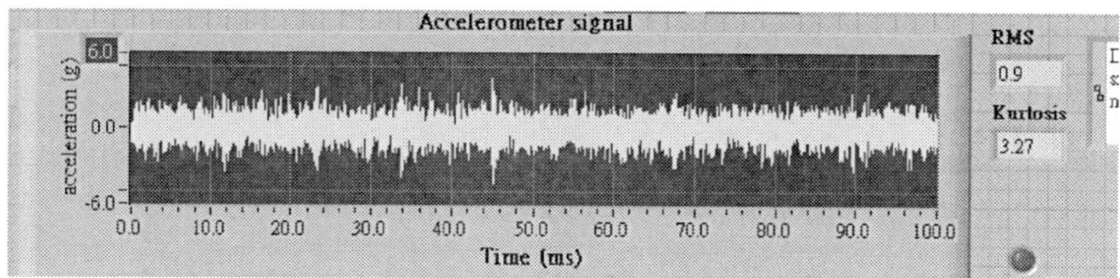

Figure 41. A combination of Gaussian white noise and outer race defective signal.

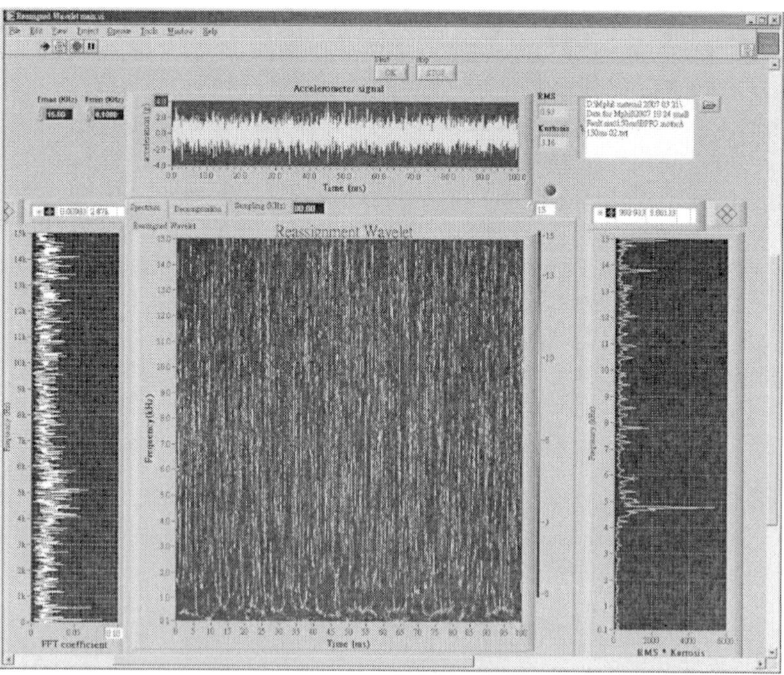

Figure 42. Reassignment Wavelet analysis for a noise submerged outer defective bearing signal.

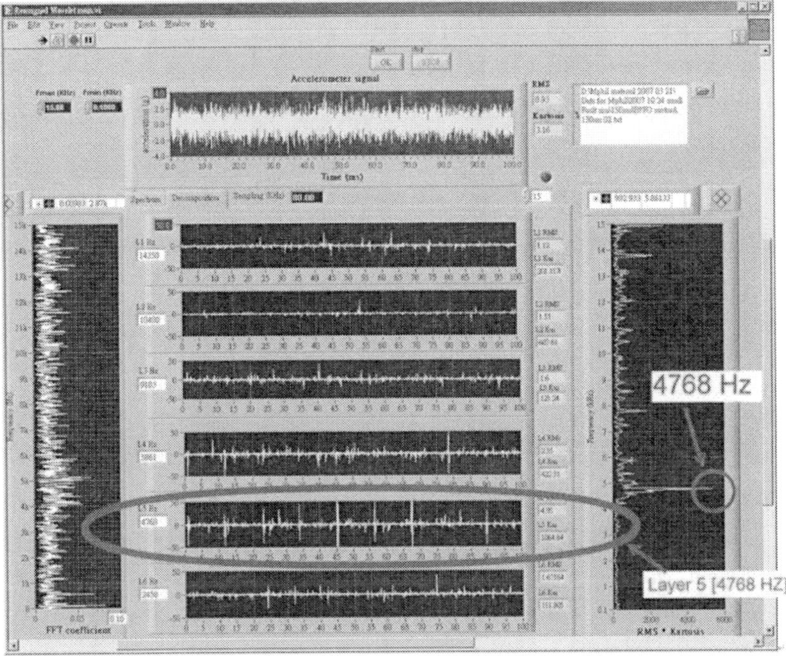

Figure 43. Reassignment Wavelet decomposition for a noise submerged outer race defective bearing signal.

Table 5. Summary of decomposition results and corresponding ranking.

Layer (Hz)	RMS	Kurtosis	RMS x Kurtosis	Rank
14350	1.12	201.1	225.23	4
10480	1.51	607.61	917.49	3
9103	1.60	123.24	197.18	5
5861	2.35	422.31	992.43	2
4768	4.95	1084.84	5369.96	1
2458	1.67	111.81	186.72	6

According to Table 5, the frequency layer of 4768Hz (highlighted red in Table 5) was ranked as number 1 in both RMS and Kurtosis values, which means that it has the largest RMS and Kurtosis values. The frequency layer of 4768 Hz (highlighted with a red circle in Figure 43) was then selected and presented in Figure 45, it was used to compare with the original data, as shown in Figure 44 and Figure 46. When comparing Figure 45 and Figure 46, impacts exist at the same time separation. To conclude, the Spectral RMS x Kurtosis can successfully detect bearing defective signals under serious noise influences.

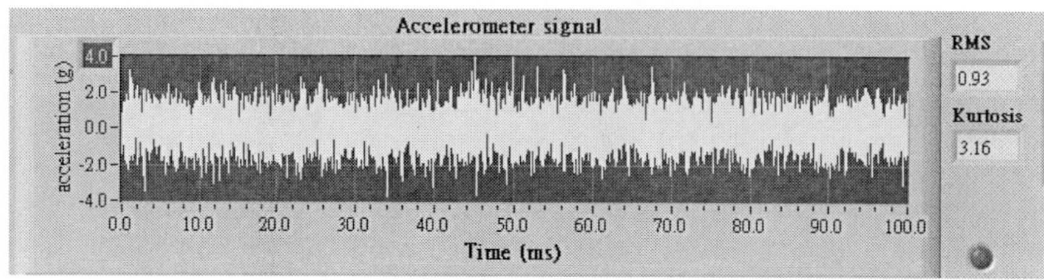

Figure 44. A combination of a Gaussian white noise and outer race defective signal.

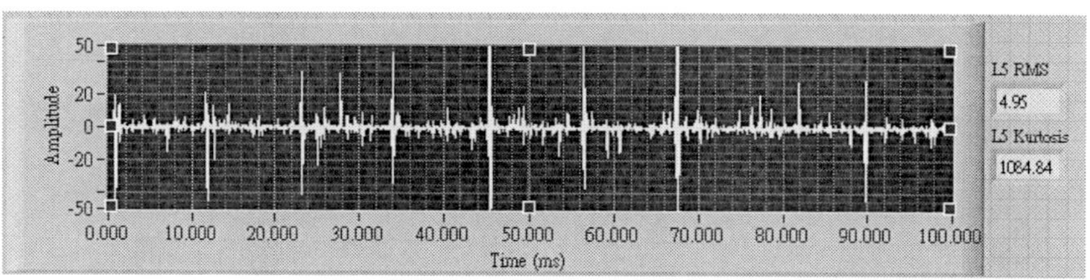

Figure 45. An extracted signal by using Reassignment Wavelet based Spectral RMS x Kurtosis (at frequency = 4786 Hz).

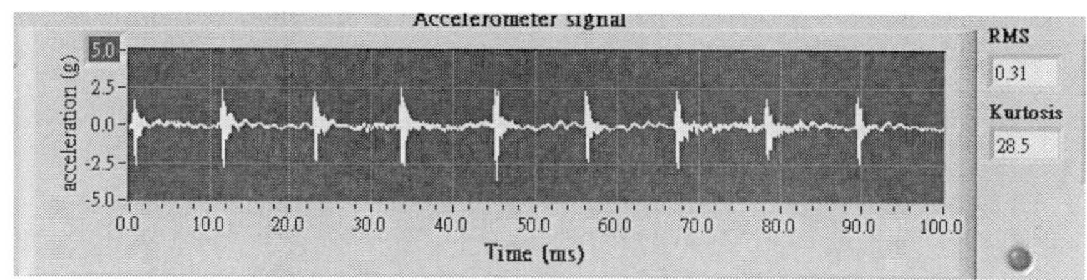

Figure 46. The original outer race defective signal.

Section 5. Experiments on Industrial Machines

After the success of the laboratory experiments, some industrial experiments were carried out to further test the new method in real situations. The first experiment was on a railway traction motor and the second experiment was on a cooling water pump.

5.1. The Analysis of the Bearing Signal Collected from a Railway Traction Motor

The Reassignment Wavelet and Spectral RMS x Kurtosis were used to analyse the vibration data of an industrial traction motor (see Figure 47). It was tested at distinct running speeds at 1498rpm (24.9Hz). The motor consisted of a 250 kg rotor which was supported by the rolling element bearings. They were located at the drive end and non-drive end. The bearing at the drive end was a single row deep groove ball bearing (SKF 6215) while that at the non-drive end was a single row cylindrical roller bearing (NU210).

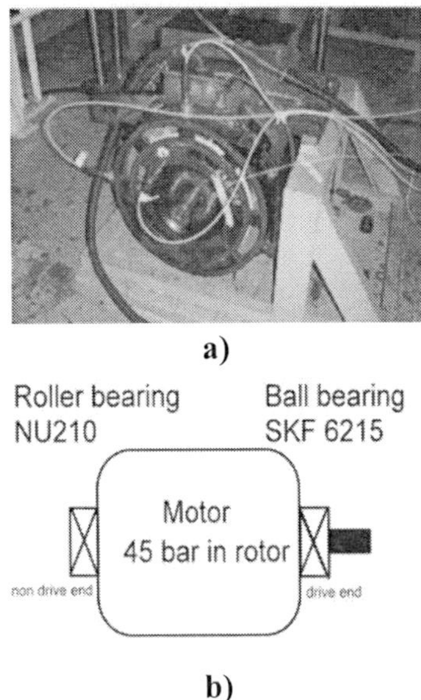

Figure 47. a) The operating industrial traction motor; b) A schematic diagram of the motor.

Table 6. The frequency order of different bearing faults

Models	SKF 6215	NU210
Outer race defect	4.57	6.86
Inner race defect	6.43	9.14
Rolling element defect	5.75	6.90

The three common types of bearing faults include outer race defects, inner race defects and rolling element defects. The frequencies of occurrence of the associate impacts, known as the bearing Characteristic Frequencies are harmonically related to the shaft speed via the frequency order they are presented in. The equation can be found previously in Section.

As mentioned earlier, in an operating machine, the rolling elements may roll over the damaged surfaces and thus produce a series of short-duration damped impacts. Theoretically, the impacts are repetitive and their frequencies of occurrence are presented in Table 7.

Table 7. Characteristic frequencies for bearing defects in a railway traction motor

Type of characteristic frequency for bearing defect	Frequency (Hz) SKF 6215	Frequency (Hz) NU210
Ball Passing Frequency Inner-race (BPFI)	114.1 Hz (8.76 ms)	171.3 Hz (5.8 ms)
Ball Passing Frequency Outer-race (BPFO)	160.5 Hz (6.23 ms)	228.2 Hz (4.4 ms)
Ball-spin Frequency (BSF)	143.6 Hz (6.97 ms)	172.3 HZ (5.8 ms)

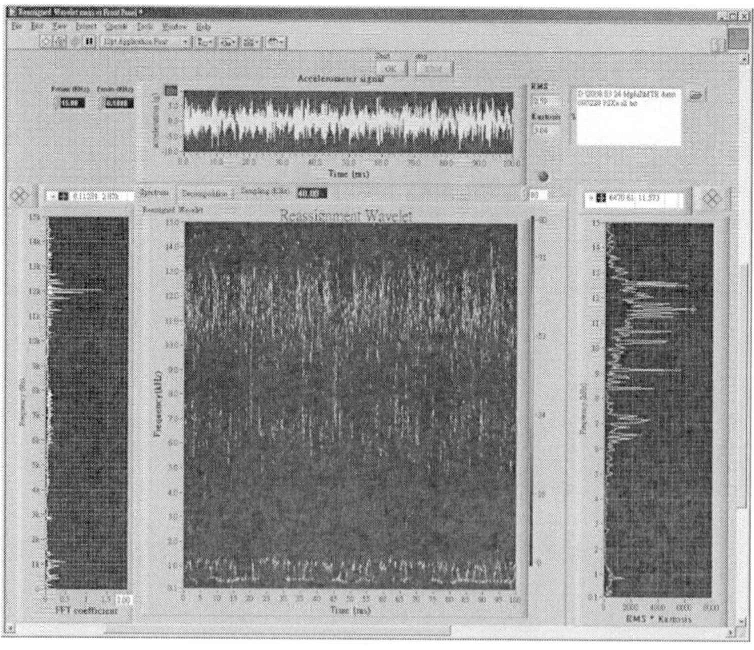

Figure 48. Reassignment Wavelet analysis on a railway industrial traction motor.

The detected bearing signal is shown in Figure 48. The left side of Figure 48 is the FFT spectrum of the inspected motor, through which it can be seen that the energy is mainly centralised in two frequency bands. The first range is located between 10 to 1000Hz and the second range located between 11.5 kHz to 12.5 kHz. To further investigate the exact frequency band of the impact, Spectral RMS x Kurtosis was used. The frequency layers of the Reassignment Wavelet were extracted and are presented in Figure 49.

Frequency layers were then ranked according to the RMS and Kurtosis values. Again, the layer with the largest value was ranked as 1 while the layer with the least value was ranked as 6. A summary of the results is shown in Table 8. The frequency layer of 11573 Hz (highlighted in red) was ranked as 1 after comparing the RMS and Kurtosis values with other frequency bands. It was then extracted and presented in Figure 50. It was used to compare with the original data, as illustrated in Figure 51.

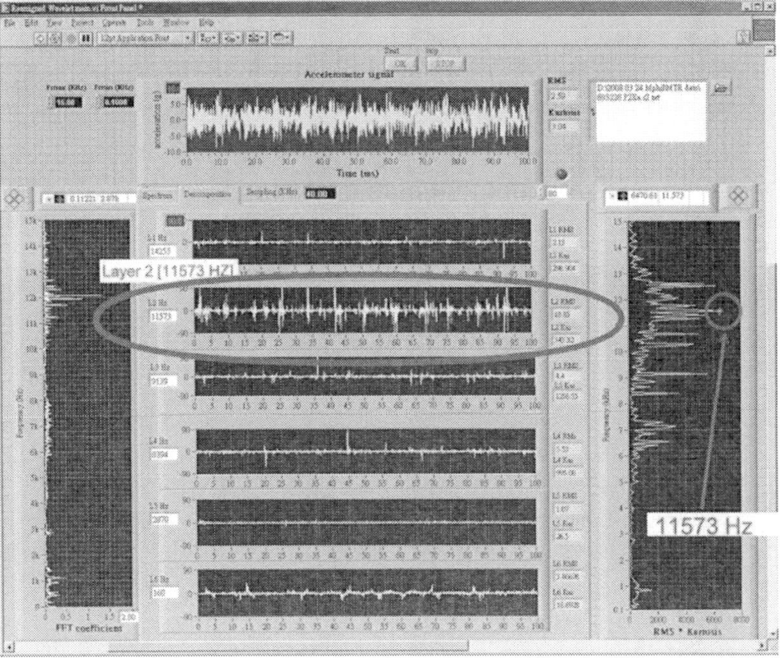

Figure 49. Reassignment Wavelet decomposition on a railway industrial traction motor.

Table 8. A summary of the decomposition results from the traction motor and the corresponding ranking

Layer (Hz)	RMS	Kurtosis	RMS x Kurtosis	Rank
14255	2.13	298.90	636.65	4
11573	18.85	343.32	6471.58	1
9139	4.4	1256.55	5528.8	2
8394	3.53	996.08	3515.8	3
2870	1.07	26.5	28.36	6
160	3.87	16.69	64.59	5

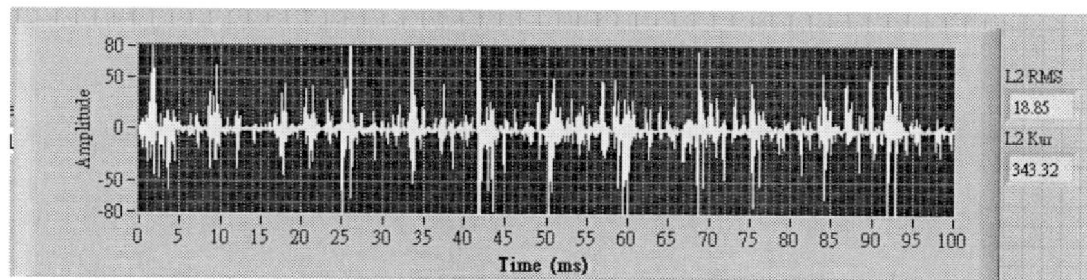

Figure 50. The extracted layer (11573 Hz) from the railway industrial traction motor.

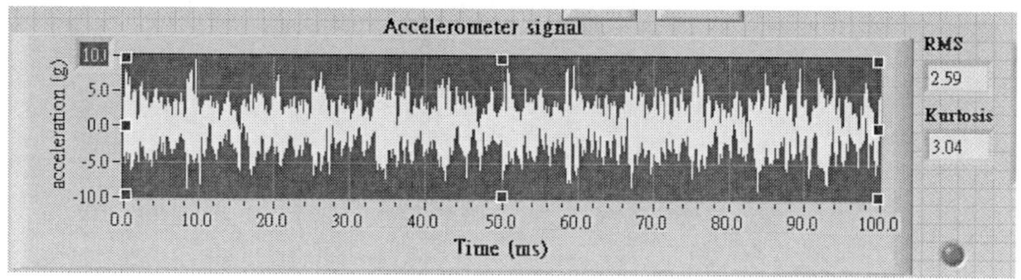

Figure 51. The original bearing signal from the railway industrial traction motor.

As shown in Figure 50, there are periodic repetitive signals captured from the running bearing. These repetitive signals were observed in previous laboratory experiences, and have been concluded to be caused by defective bearings. By measuring the time span between each interval of impacts, the period of impacts can be roughly estimated as 9 to 10ms. Since the bearing is a rolling element bearing (SKF 6215), the measured period is close to the calculated BPFI (8.76ms), which can be found in Table 7. Hence, the bearing is concluded to have become defective. Afterwards, the result was verified by the engineer of the railway company, who stated that a slight defect was found on the inner-race of the inspected bearing. The defect is shown in Figure 52.

Figure 50 shows that the repetitive impact was extracted by using the proposed method (Reassignment Wavelet based Spectrum RMS x Kurtosis), while Figure 51 shows the original signal, i.e. without the application of the proposed method, and the impact cannot be revealed clearly. This illustrates the significance of applying the newly proposed signal processing methodology.

Section 4.4 mentioned that the bearing defect inspected in the laboratory test was an artificial one and was made by a milling tool. In Figure 52, a real defective bearing is presented. The defect that existed on the bearing was not artificial. In fact, the SKF bearing is made with a very hard material. The hardness of the SKF bearing is about 63 HRC, which is harder than stainless steel. The main reason for the bearing crash was the leakage of electrical charges from the power cable that supplies electricity to the motors for driving the train. These charges leaked through one of the motors to its bearing and then tried to ground via the rail that contacted with the motor's wheel. Some of these charges failed to ground but struck the bearing. As a result, the leaking charges etched a crack on the bearing surface as shown in Figure 52. Without proper maintenance, the crack might enlarge and cause the bearing stop to move. Eventually, the entire train could stop suddenly.

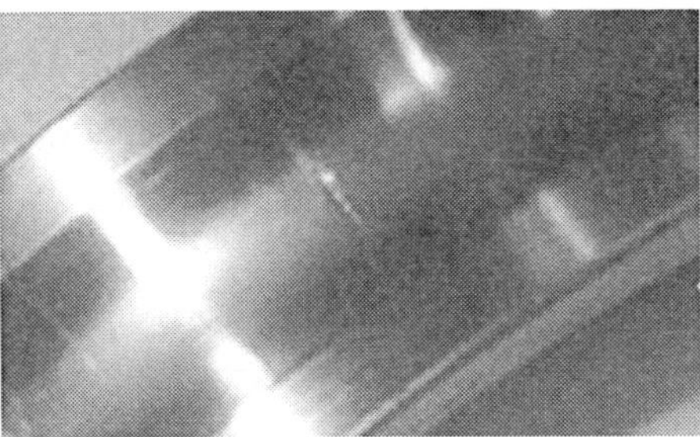

Figure 52. The defective bearing (SKF 6215).

5.2. The Industrial Cooling Water Pump

Figure 53a) shows an operating industrial cold water pump and Figure 53b) shows a schematic diagram (side view) of the pump under inspection. The driving unit is a 7.5kW 3-phase 2-pole induction motor which drives the centrifugal pump at a speed of 2900 rpm (i.e., fs = 48.33 Hz) via a universal joint. The motor consists of two single row deep groove ball bearings (Koyo 62052RS) at the non-drive end and the drive end. The centrifugal pump consists of a 5-vane impeller, which is directly connected to the universal joint at the drive end of the centrifugal pump.

Table 9. Bearing dimensions

Models	62052RS
Number of rolling elements	9
Ball diameter	10 mm
Pitch diameter	41 mm
Shaft Frequency (Fr)	2900 rpm (48.33 Hz)
Contact angle	0

Table 10. Characteristic frequencies for bearing defects in a cold water pump

Type of characteristic frequency for bearing defect	Frequency (Hz)
Ball Passing Frequency Inner-race (BPFI)	3.7 ms (270 Hz)
Ball Passing Frequency Outer-race (BPFO)	6.1 ms (164 Hz)
Ball-spin Frequency (BSF)	5.4 ms (186 Hz)
Fundamental Train Frequency (FTF)	55.6 ms (18 Hz)

Table 11. Other machinery fault identification in a cooling water pump

Other machinery fault identification	Calculation	Frequency (Hz)
Defective Motor Fan	Fr x No. of blades	338.31
Defective Pump Vane	Fr x No. of stage x No. of vanes	48.33
Unbalance	1 x Fr or 2 x Fr	48.33 or 96.66
Motor Pulsating Torque	Fr x No. of Pole x No. of phase	289.98

The previous laboratory experiment showed that at an early stage of bearing damage, a sizeable local defect may occur in one or more of the bearing components, namely, the rolling elements, the inner raceway and the outer raceway. The bearing dimension and its defective characteristic frequencies of occurrence are presented in Table 9 and Table 10, respectively. The equation can be found in Section 3.3.2. The other characteristic frequencies of machinery fault for this cooling water pump are also calculated, as illustrated in Table 11.

The bearings at the drive end and non-drive end are single-row and deep-groove ball bearings (Koyo 62052RS). The detected vibration signal is shown at the top of Figure 54. Two frequency peaks, which are highlighted with green circles on the left hand side of Figure 54, are detected on the FFT spectrum. One exists at 10316Hz, and the other one exists at 11865Hz. To investigate the exact frequency band of the impact, Spectral RMS x Kurtosis was used and only one exact freqency peak was detected at 11575 Hz (highlighted in a circle on the right hand side of Figure 54). A summary of the results is shown in Table 12. The largest RMS and Kurtosis values exist on the frequency layer of 11575 Hz.

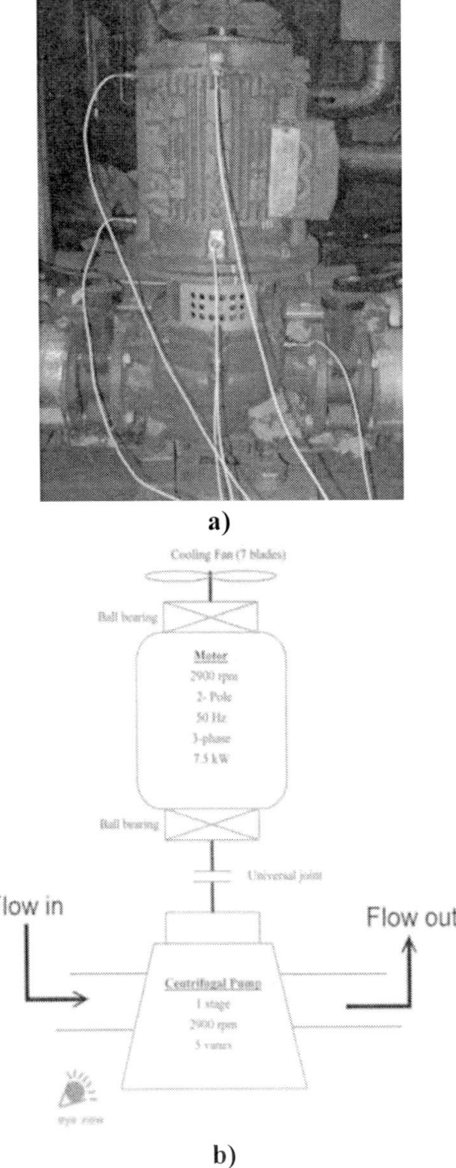

Figure 53. a) Industrial cooling water pump. b)Schematic diagram of the cooling water pump.

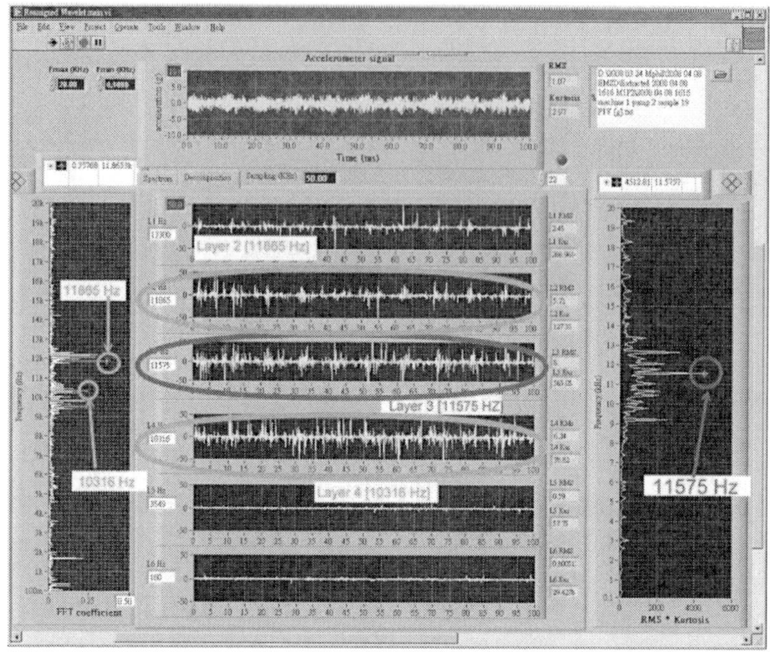

Figure 54. Reassignment decomposition in a cooling water pump.

For ease of comparison, these three selected frequency layers were extracted. The frequency layer of 11575 Hz was extracted based on the Spectral RMS x Kurtosis (as shown in Figure 55). Periodic and repetitive impacts were detected in the frequency layer of 11575 Hz. However, when the other two frequency layers of 10316Hz and 11856Hz were extracted, based on the FFT (as shown in Figure 56 and Figure 57 respectively), they were unable to illustrate the repetitive signals.

Table 12. A summary of decomposition results in a cooling water pump

Layer (Hz)	RMS	Kurtosis	RMS x Kurtosis	Rank
11865	5.71	127.35	727.17	2
11575	8	565.05	4520.4	1
10316	6.24	78.82	491.84	3

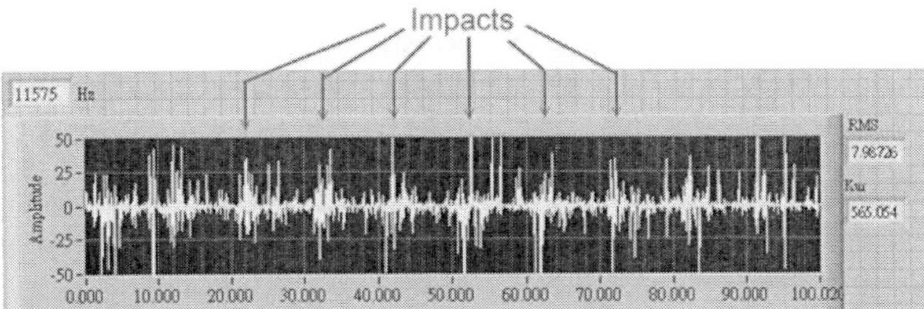

Figure 55. The extracted layer of signal collected from the cooling water pump (at frequency = 11575 Hz, the frequency peak appears on the Spectral RMS x Kurtosis).

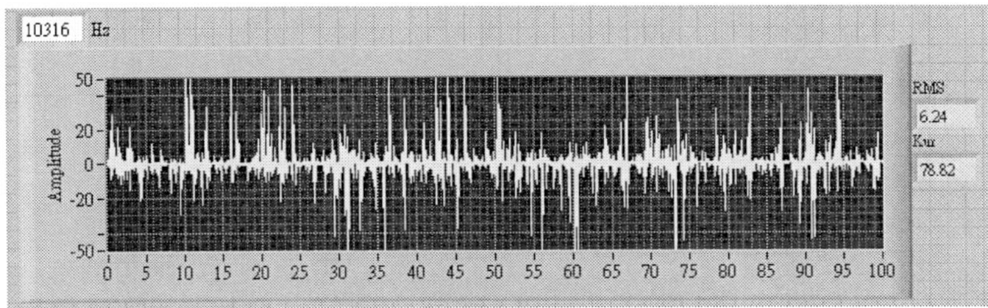

Figure 56. The extracted layer from the cooling water pump (at frequency = 10316 Hz, one frequency peak on the FFT spectrum).

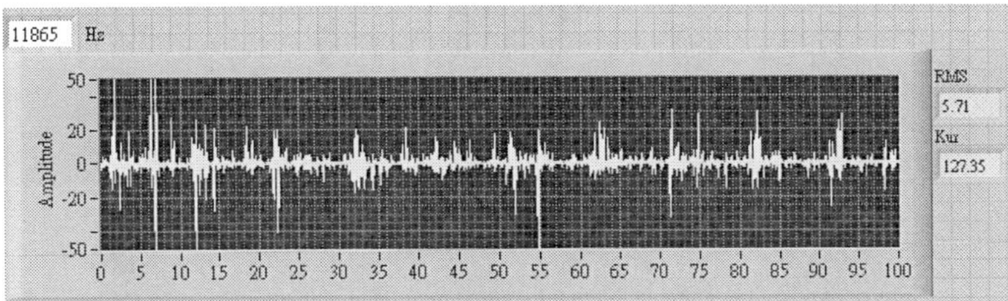

Figure 57. The extracted layer of signal collected from the cooling water pump (at frequency = 11865 Hz, another frequency peak appears on the FFT spectrum).

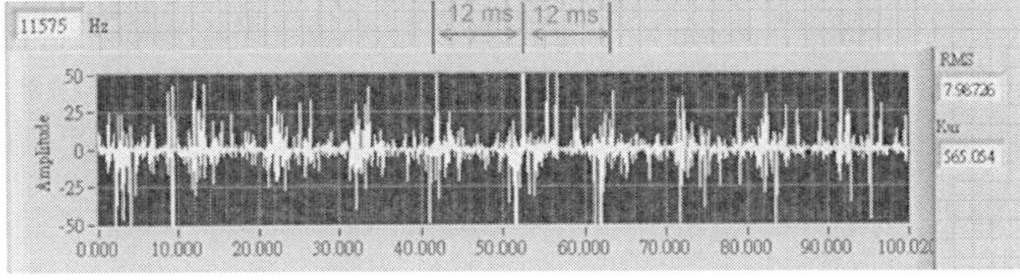

Figure 58. The extracted layer of signal collected from the cooling water pump.

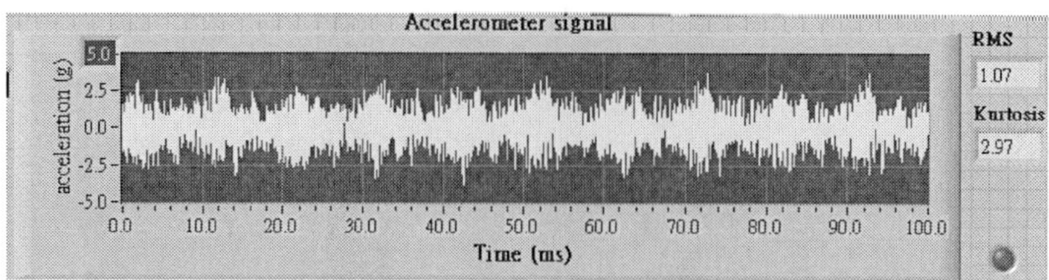

Figure 59. The original bearing signal collected from the cooling water pump.

To sum up, FFT is incapable of detecting bearing defects precisely in this situation. However, by using the Spectral RMS x Kurtosis, bearing excitation frequency can be detected successfully and the bearing defect can be revealed by extracting the frequency layer based on the detected excitation frequency.

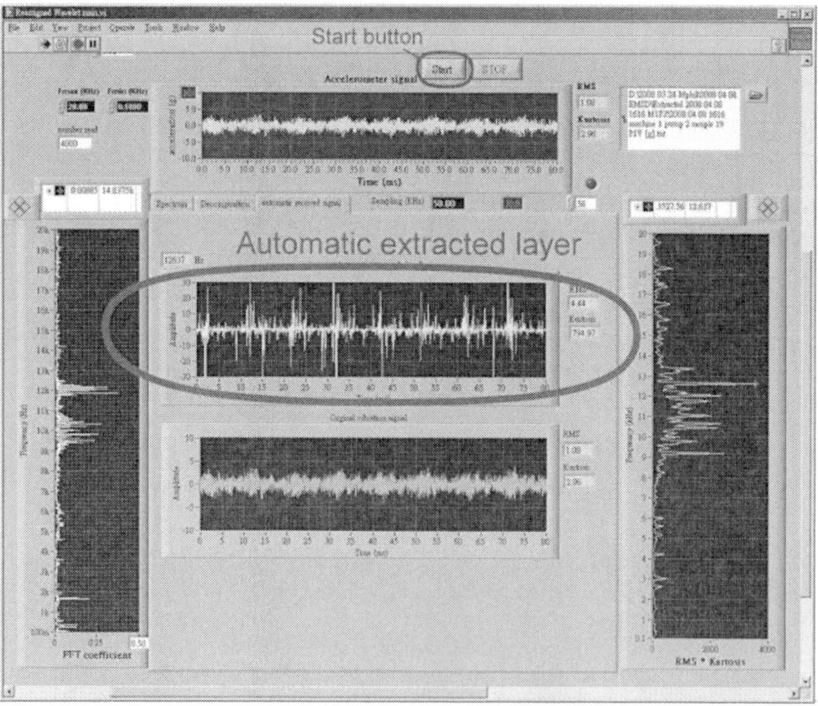

Figure 60. The front panel of the designed automatic bearing fault detection system.

Finally, by measuring the time span between the two impacts existing within the 11575 Hz layer, these impacts were concluded to be repetitive and would happen about every 12ms (as shown in Figure 58). The measured time span is nearly the twice the BPFO (2 x [6.1ms] ~ 12ms), which was calculated and can be found in Table 10. Hence, the cause of these repetitive signals being found in the layer of 11575 Hz could be concluded to be an outer race bearing defect. Also, a slight misalignment of the coupler between the motor and pump may be concluded because the detected time span was double that of the BPFO.

5.3. Automatic Signal Processing Using a Virtual Instrument

After the accomplishment of experiments on both laboratory and industrial machinery, the new methodology was incorporated into a single virtual instrument. By a single click on the 'start' button located on the front panel, as shown in the top part of Figure 60, all signal processing, e.g. data acquisition, signal analysis and graph sketching, is executed immediately and automatically. Moreover, this virtual instrument can automatically detect the frequency layer that contains the highest value on Spectral RMS x Kurtosis. Once the layer is detected, it will be extracted and displayed as a waveform diagram in the middle of the front panel, as shown in Figure 60. This virtual instrument enables a quick and fully automatic method for bearing fault detection. Use of this system can reduce maintenance costs and enhance the accuracy of fault detection, even without the existence of an expert.

Section 6. Discussion

In this chapter, the development of algorithms for rolling element bearing fault detection has been discused. Two approaches for bearing fault detection were investigated. One uses the Reassignment method to reduce the overlapping caused by the Conventional Wavelet decomposition. The other is to recover the bearing fault signals by using Spectral RMS x Kurtosis combined with Reassignment Wavelet. The experimental result shows that the Reassignment Wavelet based Spectral RMS x Kurtosis can successfully recover the bearing fault automatically. Also, they show excellent performances when extracting faulty bearing signals in a noise submerged environment.

The result generated by the traditional approach can be easily altered by interference or noise signals. In real cases, bearing signals may be influenced by ambient noise or overpowered by the nearest component when operating in an industrial plant. Focusing only on bearing characteristic frequency is not sufficient to detect a bearing defect. Therefore, a new method is necessary for recovering the faulty bearing signal precisely. The emphasis of the novel method is to detect the bearing excitation frequency zone by using higher order statistics and to recover the bearing signal based on the detected frequency.

Firstly, the Wavelet analysis was applied to decompose the bearing signal. Wavelet is conventionally used in machinery fault diagnosis to redeem the insufficient FFT in the time domain representation. However, according to the experimental results, there is a limitation included in Wavelet when it is used as a fault detection method. In Section 4.4, the results of the bearing fault simulator show that the overlapping phenomenon exists using Wavelet analysis, which makes the interpretation of the results very difficult and the recovery process does not succeed. In other words, the bearing excitation zone cannot be revealed by only using Wavelet analysis.

Secondly, the Reassignment method was investigated. The contribution of Reassignment is the increase of concentration in energy distribution. It helps to increase the readability in the time-frequency domain and provides a central location for the energy distribution in Wavelet analysis. According to our results in Section 4.4, the overlapping phenomenon in Wavelet analysis was minimised by using the Reassignment method.

The greatest problem encountered was the selection of the indictor for representing the bearing fault. Also, this indicator was used to identify the bearing excitation zone in its frequency spectrum. At this stage, some statistical indictors were investigated, including mean, peak to peak value, RMS and energy level. RMS was selected to be the fault indicator. This was because RMS is one of the adopted values for machinery fault diagnosis in the ISO standard 0 and it provides the measurement of signal intensity. Meanwhile, according to the experimental results based on the laboratory fault simulator, a bearing which begins to become defective will generate a series of repetitive impact signals. RMS may not be sufficient to reflect these short duration impulse signals. Therefore, another indictor will need to be investigated in order to identify the impulsiveness of a signal.

In this situation, Kurtosis was used. It is a statistical indicator used in the time domain. In a previous section, a comparison was made between the signals of a faulty bearing and a normal bearing. This showed that Kurtosis is a good and reliable indicator of impact. However, in most cases, when a faulty signal occurs, the Kurtosis value will go up. The detected Kurtosis value will be wrongly enlarged due to the occurrence of random small impacts caused by the noisy signal. To avoid this kind of fault alarm, we have an idea that Kurtosis be combined with the use of RMS, because RMS measures the intensity and Kurtosis indicates the impact. Finally, the spectrum of RMS x Kurtosis was introduced. It calculated the product of RMS and Kurtosis in each frequency band under Reassignment Wavelet analysis. The aim of Spectral RMS x Kurtosis is to identify the bearing excitation frequency that contains the highest opportunity to reveal the defective bearing signal.

Three stages of experiments were used to investigate the effectiveness of this new idea. First was the laboratory experiment using artificial bearing defects. Second was the stimulation using Gaussian noise on a defective bearing signal. Third was the experiment on industrial machines.

In the first stage of the experiment, three different types of bearing defects were made. The defects were fabricated by using artificially engraved bearings. The experimental results ensure that Reassignment Wavelet analysis can overcome the shortcomings of conventional Wavelet analysis. Impacts caused by defects are concentrated by using the Reassignment Wavelet. Finally, the Spectral RMS x Kurtosis was generated by calculating the product of RMS and Kurtosis in each frequency band of the Reassignment Wavelet. By using Spectral RMS x Kurtosis, the bearing excitation zone was detected and a decomposition layer of this detected frequency zone was extracted to reveal the bearing fault. The result of this experiment shows that Spectral RMS x Kurtosis can successfully reveal the excitation zones that magnify the impacts caused by a bearing defect.

In order to further investigate the effectiveness of the proposed method, the second stage of tests were done. In this stage, a Gaussian noise signal was generated to confuse the faulty bearing signal. The result is shown in Section 4.9. Although the faulty bearing signal was hidden by Gaussian noise, Spectral RMS x Kurtosis could successfully indicate the frequency layer that contains the faulty bearing features. Finally, the merged bearing signal was recovered by the detected frequency layer. A comparison between the original faulty bearing signal and the extracted bearing signal was previously presented at the end of Section 4.9. The comparisons show that the proposed approach, Spectral RMS x Kurtosis, is useful at recovering a faulty bearing signal from a noisy environment.

In the final stage, some industrial experiments were carried out to further test this novel method, Spectral RMS x Kurtosis, in real situations. The first experiment was held on a railway traction motor and the second experiment was held on a cooling water pump. In the first experiment, a sequence of periodic and repetitive signals was extracted from the original bearing signal of the railway traction motor. These repetitive signals were concluded to be generated by the bearing defect. Afterwards, the result was verified by the engineer of the railway company, who found a slight defect on the outer-race of the inspected bearing. In the second experiment, a cool water pump was tested. Two frequency layers of data with the highest peaks were extracted, based on the calculation of FFT. At the same time, a sequence of periodic repetitive signals was extracted based on the calculation of Spectral RMS x Kurtosis. The comparison of these layers showed that the frequency peaks detected in FFT are not capable of uncovering the bearing defect in this situation. However, by using the Reassignment based Spectral RMS x Kurtosis, the bearing excitation frequency could be detected and could successfully reveal the impacts caused by the bearing defect. Finally, these repetitive signals could be concluded to be caused by the bearing defect and the coupler misalignment of the cooling water pump.

To sum up, the proposed novel method of fault detection of bearings can successfully recover a faulty bearing signal from a noisy and industrial environment. It shows promising results. More importantly, this method is implemented with a virtual instrument, which enables a quick and fully automatic conversion from the original signal.

Section 7. Conclusions

In this chapter, a novel method for extracting faulty bearing signals automatically was proposed. This method was developed by incorporating the Reassignment Wavelet and the Spectral RMS x Kurtosis. Through the use of Reassignment Wavelet, the common deficiency of energy leakage that occurs in conventional Wavelets can be minimised. Meanwhile, by using Spectral RMS x Kurtosis, the excitation frequency zone of defective bearings can be determinated automatically. By applying the decomposition function of Reassignment Wavelet to this bearing excitation frequency zone, the time interval of impacts caused by bearing defect can be found. Hence, the cause of defects can be uncovered.

To prove the success of this method, extensive experiments were conducted with both laboratory machines and industrial machines. The experimental results have proved that the method is effective in detecting common bearing faults occurring in rotary machines. Both naturally developed and artificially induced defects were applied to the inspected bearings. The faulty bearing signals could still be detected accurately during the tests in spite of massive surrounding noise.

Another contribution of this chapter is the success of designing an automatic system for bearing fault detection. All of the derived processes in this system can be fully automated by implementing the novel method using a virtual instrument. It can save the analysing effort of experts and minimise the cost of maintenance.

To conclude, the novel method of combining Reassignment Wavelet and Spectral RMS x Kurtosis can help to identify the health status of bearing running in a rotary machine more easily and precisely. This method can be used to ease the workload of maintenance staff and avoid the hiring of expensive and rare experts. Moreover, a set of practical and virtual instruments were implemented successfully for automatically detecting the common faults in bearings. It provides an easy-to-use and low-cost platform for performing reliable bearing fault diagnosis.

Acknowledgments

The work that is described in this chapter was partly supported by a grant from the Innovation and Technology Commission of the Hong Kong Special Administrative Region, China (Project No. ITSP/169/01), the Research Grants Council of the Hong Kong Special Administrative Region (Project Nos. CityU 120605 and CityU 120506), and by a grant from the City University of Hong Kong, China (project no. 7002326).

References

Alfredson, R. & Mathew, J. (1985). Time domain methods for monitoring the condition or rolling element bearings. *Mechanical Engineering Transactions – Institution of Engineers*, 2, 102-107.

Ambardar, A. (1995). Analogy and digital signal processing. USA: Thomson Learning,

Antoni, J. & Randall, B. (2006). The spectral kurtosis: application to the vibratory surveillance and diagnostics of rotating machines. *Mechanical Systems and Signal Processing*, 20(2), 308-331.

Antoni, J. (2006). The spectral kurtosis: a useful tool for characterising non-stationary signals. *Mechanical Systems and Signal Processing*, 20(2), 282-307.

Auger, F. & Flandrin, P. (1995). Improving the readability of time-frequency and time-representation by the reassignment method. *IEEE Transactions on Signal Processing*, 43(5), 1068-1089.

Bakhtiary, F., Nayeb, A. & Yeganeh, S. (2005, September). Application of vibration analysis in ball bearing fault detection. *International Design Engineering Technical Conferences and Computers and Information in Engineering Conferences*, 845-855.

Bently, D., Grissom, B. & Hatch, C. (2002). Modes of vibration. *Fundamentals of rotating machinery diagnostics* (227-248). Minden: Bently Pressurized

Bearing Company.

Chui, K. (1992). Continuous wavelet transforms. *An introduction to wavelet* (7-9). Boston: Academic Press.

Dwyer, R. (1983). Use of the kurtosis statistic in the frequency domain as an aid in detecting random signals, *IEEE Journal of Oceanic Engineering*, 9(2), 85-92.

Gade, S. & Hansen, K. (1996). Non-stationary signal analysis using wavelet transform, short-time fourier Transform and wigner-ville distribution. *Bruel and Kjaer Technical Review*, 125-143.

Goldman, S. (1991). Transducers for vibration measurement. *Vibration spectrum analysis: a practical approach*, (49-77). New York: Industrial Press Inc.

Gupta, P. (1979). Dynamics of rolling element bearings part I and II. *Journal of Lubrication Technology*, 101(5), 293-318.

Habaibeh, A., Parkin, R. & Redgate, J. (2005), September). The design of enhanced condition monitoring system using sensor positioning and signal conditioning approach. *International Design Engineering Technical Conferences and Computers and Information in Engineering Conference*, 85660.

Ho, D. & Randall, B. (2000). Optimisation of bearing diagnostic techniques using simulated and actual bearing fault signals. *Mechanical Systems and Signal Processing*, 14(5), 763-788.

Inman, D. (2000). Vibration testing and experimental modal analysis. *Engineering vibration*, (364-399). Englewood Cliffs: Prentice Hall.

International Standard Organisation. (2002). *Mechanical vibration – evaluation of machine vibration by measurements on non-rotating parts*, 10816(1).

Lin, J. & Qu, L. (2000). Feature extraction based on morlet wavelet and its application for mechanical fault diagnosis. *Journal of Sound and Vibration*, 234(1), 135-148.

Kiral, Z. & Karagulle, H. (2006). Vibration analysis of rolling element bearing with defect under the action of unbalanced force. *Mechanical System and Signal Processing*, 20(8), 1967-1991.

Lindsay, J. & Asoke, K. (2000). Genetic algorithms for feature selection in machine condition monitoring with vibration signals. *Journal of Visual Image Signal Process*, 147(3), 205-212.

Larsen K. & Son A. (1989). *Machines condition monitoring*, (20-23). Denmark: Bruel and Kjaer.

Larsen K. & Son A. (1989). *Machine health monitoring using vibration analysis*. (15-20). Denmark: Bruel and Kjaer.

Lee, J. & Scott, L. (2006, July). Zero-breakdown machines and systems productivity needs for next-generation maintenance. *World Congress of Engineering Asset Management*, 100-108.

Leung, J. & Tse, P. (2005). An economical engineering monitoring system with embedded advanced fault diagnostic features. *International Journal of Industrial Engineering Theory and Application*, 4(1), 38-50.

Li, C. & Ma, J. (1997). Wavelet decomposition of vibrations for detection of bearing-localized defects. *NDT and E International*, 30(5), 143-149.

Li, L. & Qu, L. (2003). Cyclic statistics in rolling bearing diagnosis. *Journal of Sound and Vibration*, 267(5), 253-265.

Lorenzo, F. & Calabro, M. (2007, June). Kurtosis: a statistical approach to identify defects in rolling bearing. *International Conference on Marine Research and Transportation*, 607-610.

Mathew, J. (1987). The condition monitoring of rolling element bearings using vibration analysis. *Journal of Vibration, Acoustics, Stress and Reliability in Design*, 106(5), 447-453.

McFadden, P. & Smith, J. (1985). The vibration produced by multiple point defects in rolling element bearing. *Journal of Sound and Vibration*, 98(2), 263-273.

McFadden, P. & Smith, J. (1984). Vibration monitoring of rolling element bearings by the high frequency resonance technique – a review. *Tribology International*, 17(1), 1-10.

Mechefske, C. (1999). Machine condition monitoring and fault diagnostics. *Vibration and Shock Handbook*, (279-289). USA: CRC Press.

Neale, N. (1980). Condition monitoring methods and their interpretation. *A guide to the condition monitoring of machines* (pp.50-89). London: Department of Trade and Industry Press.

Newland, D. (1994). Wavelet analysis of vibration - part 1: theory and part 2: wavelet Maps. *Journal of Vibration and Acoustic*, 116(5), 409-425.

Nyquist, H. (1924). Certain factors affecting telegraph speed. *Bell System Technical Journal*, 3(5), 324-346.

Ohta, H. & Sggimoto, N. (1996). Vibration characteristics of tapered roller bearing. *Journal of Sound and Vibration*, 190(2), 137-147.

Peled, R., Braun, S. & Zacksenhouse, M. (2005). A blind de-convolution separation of multiple sources with application to bearing diagnostics. *Mechanical Systems and Signal Processing*, 19(6), 1181-1195.

Peng, Z., Chu, F. & He, Y. (2002). Vibration signal analysis and feature extraction based on reassignment wavelet scalogram. *Journal of Sound and Vibration*, 253(5), 1087-1100.

Petrosian, A. & Meyer, F. (2001). Multi-scale analysis, estimation and filtering.

Wavelet in signal and image analysis, (101-130). Netherland: Kluwer Academic Publisher.

Ramesh, T., Srikanth, S. & Sekhar, A. (2008). Hilbert–huang transform for detection and monitoring of crack in a transient rotor. *Mechanical Systems and Signal Processing*, 22(4), 905-914.

Randall, B. (1987). Computation using fast fourier transform. *Frequency analysis*, (50-65). Denmark: Bruel and Kjaer.

Sawalhi, N. & Randall, B. (2006). Helicopter gearbox bearing blind fault identification using a range of analysis techniques. *Australian Journal of Mechanical Engineering*, 5(2), 157-168.

Sawalhi, N., Randall, B. & Endo, H. (2007). The enhancement of fault detection and diagnosis in rolling element bearings using minimum entropy de-convolution combined with spectral kurtosis. *Mechanical Systems and Signal Processing*, 21(6), 2616-2633.

Sadettin, O., Nizami, A. & Veli, C. (2007). Vibration monitoring for defect diagnosis or rolling element bearings as a predictive maintenance tool: comprehensive case studies. *NDT & E International*, 39(4), 293-298.

Stein, M. & Shakarchi, R. (2003). Convergence of fourier series. *Fourier analysis-an introduction*, (69-99). New Jersey: Princeton University Press.

Strang, G. & Nguyen, T. (1996). Wavelet theory. *Wavelets and filter banks*, (221-259). Wellesley: Wellesley Cambridge Press.

Sandy, J. & Harker, R. (1989). Rolling element bearings monitoring and diagnostics techniques. *Journal of Gas Turbines and Power*, 111(5), 251-256.

Tang, S. (2002). On the addition of two incoherent unsteady noise records of similar statistical structures, *Journal of Applied Acoustics*, 63(8), 829-848

Tandon, N. & Choudhury, A. (1999). A review of vibration and acoustic measurement methods for the detection of defects in rolling element bearings. *Tribology International*, 32(8), 469-480.

Tao, B., Zhu L., Han, D. & Xiong, Y. (2007). An alternative time-domain index for condition monitoring of rolling element bearings – a comparison study. *Journal of Reliability Engineering and System Safety*, 92(5), 660-670.

Tse, P., Gontarz, S. & Wang, X. (2006). Enhanced eigenvector algorithm for recovering multiple sources of vibration signals in machine fault diagnosis. *Mechanical Systems and Signal Processing*, 21(7), 2794-2813.

Tse, P. & Lai, W. (2003, August). Minimizing interference in bearing condition monitoring and diagnosis by using adaptive noise cancellation. *International Congress and Exhibition on Condition Monitoring and Diagnostic Engineering Management*, 739-748.

Tse, P., Peng, Y. & Yam, R. (2001). Wavelet analysis and envelope detection for rolling element bearing fault diagnosis – their effectiveness and flexibilities. *Journal of Vibration and Acoustics*, 123(3), 303-310.

Tse, P. & Yang, W. (2002, July). A new wavelet transform for eliminating problems usually occurring in conventional wavelet transforms used for fault diagnosis. *The Ninth International Congress on Sound and Vibration*, 465-473.

Tse, P. & Yang, W. (2003). Development of an advanced noise reduction method for vibration analysis based on singular value decomposition. *NDT & E International*, 36(6), 419-432.

Tse, P. & Yang, W. (2002, September). The practical use of wavelet transforms and their limitations in machine fault diagnosis. *International Symposium on Machine Condition Monitoring and Diagnosis*, 9-16.

Tse, P., Yang, W. & Tam H. (2004). Machine fault diagnosis through an effective exact wavelet analysis. *Journal of Sound and Vibration*, 277(4-5), 1005-1024.

Wang, J., Tse, P., He, L. & Yeung, R. (2004). Remote sensing, diagnosis and collaborative maintenance with web-enabled virtual instruments and mini-servers. *International Journal of Advanced Manufacturing Technology*, 24(9-10), 764-772.

Wang, J. & McFadden, P. (1996). Application of wavelet to gearbox vibration signals for fault detection. *Journal of Sound and Vibration*, 192(5), 927-939.

Wang, L., Xu, G., Wang, J., Yang, S. & Yan, W. (2008, July). Application of hilbert-huang transforms for the study of motor imagery tasks. *Engineering in Medicine and Biology Society*, 3848-3851.

Wowk, V. (1991). Machinery monitoring. *Machinery Vibration- Measurement and Analysis*, (17-18). New York: McGraw-Hill Inc.

Yan, R. & Gao, R. (2006). Hilbert–huang transform based vibration signal analysis for machine health monitoring. *IEEE Transactions on Instrumentation and Measurement*, 55(6), 2320-2329.

In: Fault Detection Theory, Methods and Systems ISBN: 978-1-61728-291-1
Editor: Léa M. Simon, pp. 69-95

Chapter 2

MONITORING AND DIAGNOSIS OF DISCRETE EVENT SYSTEMS USING TIME PETRI NETS, A RAILWAY CASE STUDY

Mohamed Ghazel[1,2*]
[1]Univ Lille Nord de France F-59000 Lille
[2]INRETS-ESTAS 20 rue Elisée
Reclus F-59666 Villeneuve d'Ascq

Abstract

Monitoring of discrete event systems has been the subject of several studies during the last decades. The approaches developed are different as regards the model used, the hypothesis taken and the point of view considered. In this chapter, a monitoring approach for timed discrete event systems is discussed. The system behavior is modeled using Time Petri Net. The proposed approach exploits temporal constraints to assess the system state and therefore detect and identify faults being given the partial observability of events. This approach uses an offline synthesis of the timed system behavior which serves as a basis to the online monitoring. Then, an online algorithm is used to track the system's state and to identify the event scenarios which occur within the system. Several mechanisms have been elaborated in order to detect the faulty behaviors and to identify the possible failures. They are based both on the structural analysis of the system behavior and on the time constraints. Moreover, the

*E-mail address: mohamed.ghazel@inrets.fr

developed approach offers the ability to anticipate the occurrence of some failures which is an interesting fact, especially when dealing with critical systems. Diagnosability questions are also put forth and some results are proposed. A railway control case study is used in order to gradually illustrate the various approach steps.

Keywords: Monitoring, Diagnosis, Discrete event systems, Online tracking, State observer, Time Petri net, Partial observability.

1. Introduction

The monitoring activity becomes more important within automated systems. Such systems, indeed, cannot be efficient without implementing reliable means for their monitoring, which are able to detect, identify and handle the possible failures. This is as much true as failures may have serious consequences like in critical systems (transport, energy, etc.) where the incurred risks are both human and material. On the other hand, it has been proved that temporal aspects can carry crucial information for the monitoring process [6]: the fact that a given event e occurs at a given date t_1 could have a very different significance, from the monitoring point of view, than if e occurs at a different date t_2. In this chapter, an online monitoring approach for timed systems is proposed. As input, it has the behavioural model of the system in the shape of a Time Petri Net (TPN),as well as the observable events caught online. Using TPN makes it possible to specify temporal uncertainty and tolerance on the system's behaviour. The underlined idea of the mechanisms developed within our approach is to analyze the timed signatures of the system's behaviour in order to detect the possible abnormalities and to identify the failures which may have occurred. Another important facility provided is the anticipation of failures' occurrence. As will be shown thereafter, this is typically a result which would not be obtained with untimed analysis. Anticipating failures is crucial especially when dealing with critical systems. The chapter is organized as follows: we give an overview on the related work in section 2. In section 3, a brief description of the considered case study is provided. In section 4, the foundations of our approach are discussed gradually when exposing the various concepts elaborated. Some resylts dealing with the In section 5, we first develop a monitoring tool -the Estimator- which offers a synthesis of the system's behaviour. An algorithm for the Estimator building is proposed. Then, an illustration on our case study is performed. diagnosability are exposed in section 6 and in section 7, the online monitoring process is explained and an algorithm is proposed to implement the elaborated mechanisms. Finally, we conclude the study while sketching some future work.

2. Related Work

In the literature, several works dealing with monitoring of DES under partial observability can be found. They are different mainly regarding both the models used and the techniques deployed. Comparatively, the main originality of the work discussed here consists, first, in using time information to make the state tracking process more effective and to obtain finer monitoring results. Then, under certain conditions, the mechanisms we developed makes it possible to predict the system behavior. When we consider the approach of Lafortune and Sampath [8], decisions are only made after an observable event occurs and event scenarios cannot be foreseen using temporal data since the model used is not timed. The "prediction" capacity offered by the approach discussed here is very important especially in critical systems with serious failure consequences like in transportation systems. In [10], Pandalai and Holloway propose templates generation for the online distributed monitoring of Dynamic DES starting from timed automata specification. In this approach, timing and sequencing relationships are used to detect abnormal behaviors (unexpected observation). However the monitoring result only declares an abnormal (faulty) behavior and does not identify what happened in the system following an unexpected observation. Another interesting approach dealing with state estimation of DES is that of Zad and Wonham [15]. In this approach time factor is considered as clock ticks whereas we consider (almost) continuous time in our approach (values in $\mathbb{Q}$). Using clock ticks considerably affects the computation complexity as one may be constrained to choose small time periods in order to express time constraints. Finally, we have to mention the work of Tripakis [13] on the diagnosis of DES under a dense-time specification. This work also focuses on diagnosability issues and defines necessary and sufficient conditions for it. The specification model used is dense-time automata.

3. The Considered Case Study

The chosen case study is introduced below and will constitute a filigree throughout the chapter. The developed monitoring means will be illustrated progressively on the basis of this case study.

The case study discussed here is inspired from a railway maquette on which we have developed some research work (figure 1). The trains are controlled with some PLCs which can communicate through a Fipway network in order to manage shared resources (track section, switches). Trains' movements are controlled by electrifying the rail segments (track sections) and by positioning the switches. Some track sections are uninterruptedly powered in such a way that when a train reaches them, it cannot stop before reaching the next section. Switches are controlled with bistable positioning orders: every switch A_i can be set direct (A_id) or bifurcated (A_ib). Since several trains may circulate simultaneously on the circuit and in order to insure an optimal operation of the railway, positioning orders are sent to the switches as late as possible before the arrival of the pertinent trains.

Moreover, for monitoring purposes, the circuit is fitted with some digital sensors set up on some segments. A sensor detects the passage of a train on the associated segment. The sensors' signals are the only external events that can be picked up online. For the sake of simplicity, we consider that we have to control -and monitor- one train using one PLC.

In figure 1, a picture of the railway circuit is shown and the considered part of the circuit is indicated. Here T_i's correspond to track sections, A_j's to switches and S_k's to sensors.

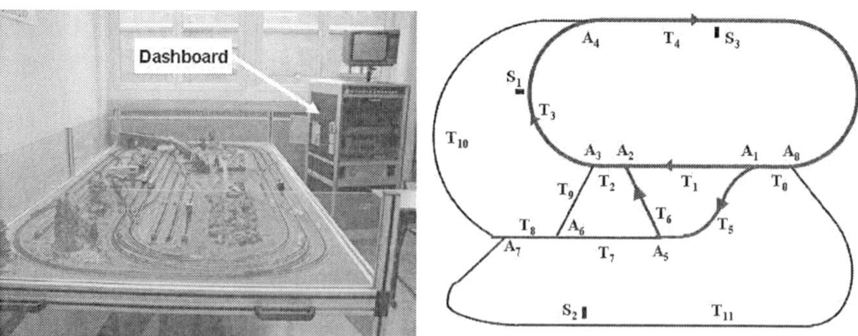

Figure 1. The Railway Maquette and the considered circuit part.

The "normal" cycle of the train is the following: $(T_0.[T_1 \text{ OR } T_5.T_6].T_2.T_3.T_4)^*$. Here, for maintenance reasons, we allow the train to go over T_1 or $T_5.T_6$. The route choice is performed by the control part (PLC). We consider, moreover, that two potential faults may occur: the first, denoted D_1, when switch A_3 is blocked in the bifurcated position, and the second, denoted D_2, when switch A_5 remains blocked in the direct position. As can be seen on the circuit, if D_1 or D_2 occurs, the train is brought towards T_8, then, according to A_7's position, towards T_{10} or T_{11}. Thus, the train itinerary becomes the following: $T_0.(T_5.T_7 \text{ OR } T_1.T_2.T_9).T_8.(T_{10}.T_4 \text{ OR } T_{11}).T_0$. Here A_5 and A_6 are controlled with the same system (both direct or both bifurcated), A_4 is set direct and the train control is set up in such a way that A_8 is positioned according to whether or not sensors S_2 and S_3 are triggered. No restriction is applied on A_7 (unknown default position); this explains the $(T_{10}.T_4 \text{ OR } T_{11})$ alternative.
The behavioral model of our system is represented in figure 2. On this TPN, transitions stand for event occurrences. Events may be observable or unobservable. Here, only events relating to the monitoring process are taken into account. Moreover, PLC orders are not represented in this model, that is to assume that the PLC operation is safe.

Let us assume that T_o denotes the set of transitions corresponding to observable events and T_{uo} the set of transitions associated to unobservable events. Here the white transitions are those in T_o and the grey ones are those in T_{uo}. Table 1 summarizes the events associated to the transitions in the TPN of fig. 2. In this table, each event A_id (respectively A_ib) represents the train passage on switch A_i in the direct (respectively bifurcated) direction.

Thus, $T_o = \{t_4, t_6, t_{16}\}$ and $T_{uo} = T \setminus T_o$ where $T = \{t_i, i \in [\![1, 19]\!]\}$.

Without loss of generality, each transition in the TPN model of Fig. 2 corresponds to a different event. This is not mandatory in our approach.

4. Monitoring Approach - Principle, Foundations

4.1. Introduction

As for the main existing monitoring approaches, our approach is model-based. The model used is a TPN [9] which gives a complete description of the system's behaviour as timed sequences of events. Concretely, in our model, transitions correspond to events' occurrences. The choice of TPN is justified by its rich semantics and by its capacities of expression. In our model, events can be of two types: observable (belonging to Σ_o) or unobservable (belonging to Σ_{uo}). In particular, faults correspond to unobservable events; the monitoring of observable faults being obvious and is reduced to an observation process. The confronted challenge is thus: how to effectively use the knowledge one has of the system's behaviour (behavioural model known a pri-

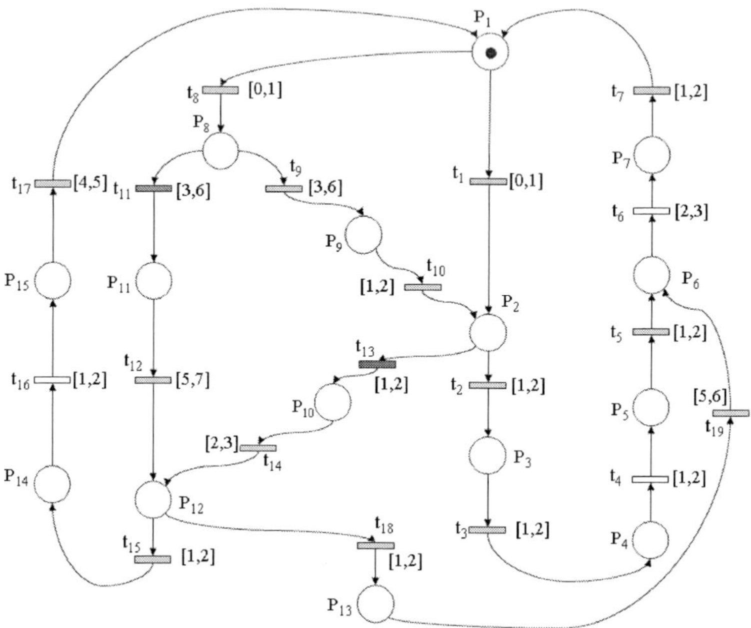

Figure 2. The Behavioral model of the system.

ori) and the observations gathered online in order to insure the system's monitoring (detection/diagnosis). The developed mechanisms are indeed not limited to the detection activity (determine whether the system behavior is normal, i.e that no fault has occurred). In addition, such mechanisms strive to determine precisely the type of the possible faults (identification/diagnosis).

Our approach suggests the building of a State Estimator starting from the state graph of the system. The establishment of the Estimator is done according to an algorithm that we developed in [6] and which will be briefly discussed in section 5.. Basically, a state class of a TPN represents a set of states of the net which can be obtained the one from the others by a simple time translation. State class notion is similar to the *state region* notion for Timed Automata [1]. The set of state classes of a TPN is called the *Class Graph* and may be finite or infinite. The readers who are not familiar with TPN and with the notion of state class can refer to [5], [6] and [7].

The Estimator is a sort of SFA which stands for a determinization of the system's behaviour depicted in the class graph. Thus, the global structure of the Estimator is obtained by an $\epsilon-$reduction on the class graph of the TPN model. Then, the **Nodes** in

Table 1. Transition-Event Correspondences

Tr	Evt	Obs.	Signification
t_1	A_1d	unobs	train on switch A1 direct
t_2	A_2d	unobs	train on switch A2 direct
t_3	A_3d	unobs	train on switch A3 direct
t_4	$\uparrow S1$	obs	sensor $S1$ triggered
t_5	A_4b	unobs	train on switch A4 bifurcated
t_6	$\uparrow S3$	obs	sensor $S3$ triggered
t_7	A_8d	unobs	train on switch A8 direct
t_8	A_1b	unobs	train on switch A1 bifurcated
t_9	A_5b	unobs	train on switch A5 bifurcated
t_{10}	A_2b	unobs	train on switch A2 bifurcated
t_{11}	A_5d	unobs	train on switch $A5$ direct
t_{12}	A_6d	unobs	train on switch $A6$ direct
t_{13}	A_3b	unobs	train on switch A3 bifurcated
t_{14}	A_6b	unobs	train on switch A6 bifurcated
t_{15}	A_7b	unobs	train on switch A7 bifurcated
t_{16}	$\uparrow S2$	obs	sensor $S2$ triggered
t_{17}	A_8b	unobs	train on switch A8 bifurcated
t_{18}	A_7d	unobs	train on switch A7 direct
t_{19}	A_4d	unobs	train on switch A4 direct

this model contain state classes and the arcs between nodes are labeled with observable events. Similar mechanisms have already been elaborated in the SFA (state finite automata) field by Lafortune and Sampath [8] who develop **Diagnosers** for DESs starting from event models. Ushio et al. [14] propose also to build **Observers** based on PN models in order to monitor DESs. Here, we integrate, in addition, temporal aspects. Structuring the set of classes in nodes allows us to "abstract" the indeterminism generated by the partial observability on the system's behavior; indeed one can determine online, and precisely, which node the system's state is in (cf. Proposition 1 in the following).

Here, the classes of each obtained node N are divided into two subsets: the set of **Entry classes** (SEC) which may be reached immediately after N is entered and the set of **Shadow classes** (SSC) which can be reached starting from a given class in SEC upon an unobservable sequence. Moreover, the temporal information about

the system's behaviour (model) is investigated in order to prepare a foundation for the online monitoring process. Hence, entry classes are enriched with **candidate sequences** and **previous sequences**.
All these new notions will be presented in the following section.

4.2. Terminology and Foundations

4.2.1. Theorem

The number of classes of a Time Petri Net is finite iff this net is bounded. $\Rightarrow$ is obvious and the proof of $\Leftarrow$ can be found in [5], as well as several sufficient conditions for the *bounded.*

4.2.2. Mappings

- The mapping evt which associates every transition with its corresponding event: $evt : T \longrightarrow \Sigma = \Sigma_o \cup \Sigma_{uo}$
  Σ is the set of events, Σ_o is the set of observable events and Σ_{uo} is the set of unobservable events.
  evt is not necessarily injective: $evt(t_i) = evt(t_j) \nRightarrow t_i = t_j$, that means two different transitions can correspond to the same event.

- The mapping tr which attributes to every event in Σ the set of corresponding transitions. In other terms, $tr = evt^{-1} : \Sigma \longrightarrow P(T)$
  $P(T)$ being the partitions' set of T.
  For $e \in \Sigma$ and $t \in T$, $t \in tr(e)$ iff $evt(t) = e$.

- The transition mapping f which manages the transitions between classes

$$f : \mathcal{C} \times T^* \longrightarrow \mathcal{C}$$

  Let c_i and c_j be two classes in $\mathcal{C}$, t a transition in T and s a sequence in T^*.
  $f(c_i, s) = c_j$ iff s connects c_i to c_j in the graph of classes. This mapping verifies: $f(x, st) = f(f(x, s), t)$.

- The unobservable reachability mapping ($\mathcal{UR}$) which enables to find the set of classes reachable from a given class c following the firing of all unobservable sequences (sequences of T^*_{uo}) which are achievable from c: $\mathcal{UR} : \mathcal{C} \longrightarrow P(\mathcal{C})$
  $c_i \longmapsto \mathcal{UR}(c_i) = \{c_j \in \mathcal{C} / \exists\ s \in T^*_{uo}, f(c_i, s) = c_j\}$
  where $P(\mathcal{C})$ is the set of partitions of $\mathcal{C}$.

- The transition mapping f_N between the Estimator nodes. Let us denote $\mathcal{N}$ the Estimator nodes set.

$$f_N : \mathcal{N} \times \Sigma_o \longrightarrow \mathcal{N}$$

for $N_i, N_j \in \mathcal{N}$, $f_N(N_i, e) = N_j$ iff there is $t \in tr(e)$, c_i a class of N_i, c_j a class of N_j and $s \in T_{uo}^* \cup \emptyset$ such that: $f(c_i, s.t) = c_j$
In an informal manner, it means that the node N_j is a successor of the node N_i in the Estimator, and that online, the transition $N_i \longrightarrow N_j$ is the consequence of the detection of the observable event e.
Conversely, if for a given node $N_i \in \mathcal{N}$ and a given event $e \in \Sigma_o$ we have: $\forall\, c_i$ of N_i, $\nexists\,(t, c_j) \in tr(e) \times \mathcal{C}$ such that: $f(c_i, t) = c_j$, then $f_N(N_i, e) = \emptyset$.

- The mapping $succ$ defined on the set $\mathcal{N}$. This mapping determines, for a given node N the set of successor nodes in the Estimator:
  $succ : \mathcal{N} \longrightarrow P(\mathcal{N})$
  $\quad N_i \longmapsto \{N_j \in \mathcal{N} / \exists\, e \in \Sigma_o, f_N(N_i, e) = N_j\}$
  In that way, one can define for every node $N_i \in \mathcal{N}$, the set of successor nodes with the corresponding observable events:
  $NEXT(N_i) = \{(N_j, e_k) \in \mathcal{N} \times \Sigma_o,\ f_N(N_i, e_k) = N_j\}$

4.2.3. Concepts and Notations

- Given a class $c \in \mathcal{C}$, we shall call **candidate sequence** of c any sequence $s.t$, $s \in T_{uo}^* \cup \emptyset$ and $t \in T_o$, which is achievable starting from c. Factual scenarios which correspond respectively to these sequences are said **candidate scenarios**. For each candidate sequence $s.t$, one associates a relative framing to its duration. Let us denote $d_{\max}(s.t)$ the upper bound of the duration interval of sequence $s.t$. Moreover, one determines for each of candidate sequences its **destination class** c', which is the class reached following the achievement of $s.t$ from c, $c' = f(c, s.t)$.
  Let us note that the candidate sequences of entry classes in a given node are used online only when the system's state is in this node.
  While online, the fact of exceeding the relative date $d_{\max}(s.t)$ (starting from the occurrence date of the last observable event -or from zero absolute date if no observable event has still occurred) without any observable event having occurred- points out that the candidate scenario corresponding to sequence $s.t$ cannot occur any more. We say in this case that sequence $s.t$ becomes **impossible** and that candidate scenario corresponding to this sequence becomes **impossible**. Before the relative date $d_{\max}(s.t)$, and always in the same circumstances (no observable event occurring), we say that the candidate sequence $s.t$ is **possible** and that the corresponding candidate scenario is also **possible**.

- Let us consider a node N_j in the Estimator and $c_j \in SEC(N_j)$. A **previous sequence** of c_j is every sequence $s.t$, $s \in T_{uo}^* \cup \emptyset$ and $t \in T_o$, such that there exists a class c_i in the set of entry classes of a node N_i predecessor of N_j, proving: $f(c_i, s.t) = c_j$. To every previous sequence, one associates its duration's **interval** and its **origin class**. For example, here the origin class of the previous

sequence $s.t$ of c_j in $SEC(N_j)$ is c_i. Similarly, factual scenarios relative to previous sequences are called **previous scenarios**. In a similarly way to candidate sequences (and scenarios), previous sequences (and scenarios) can be in the state **possible** or **impossible** (see monitoring algorithm, section 7.2.).
The calculation of a sequence's duration is done according to a linear algebra technique that can be found in [5].

- Given a node N of the Estimator, $c \in SEC(N)$ and $t \in T_o$, we denote:
  - SCS^c (Set of Candidate Sequences of class c): the set of sequences $s.t$, $s \in T^*_{uo} \cup \emptyset$ and $t \in T_o$ which are achievable from c. Formally we can write:
    $SCS^c = \{s.t,\ s \in T^*_{uo} \cup \emptyset,\ t \in T_o \text{ and } f(c, s.t) \neq \emptyset\}$
  - SCS_N (Set of Candidate Sequences of node N): the set of sequences $s.t$, where $s \in T^*_{uo} \cup \emptyset$ and $t \in T_o$ which are achievable from all entry classes in N, i.e:
    $$SCS_N = \bigcup_{c_i \in SEC(N)} SCS^{c_i}$$
  - SCS^t_N: the set of all sequences $s_j.t$, $s_j \in T^*_{uo} \cup \emptyset$ which are achievable from all entry classes in the node N. Formally SCS^t_N can be written as:
    $$SCS^t_N = \bigcup_i \{s_i.t \in SCS_N\}$$
  - $T_{N-obs-next}$: the set of transitions corresponding to the observable events which can occur starting from the classes of the node N.
    $T_{N-obs-next} = \{t \in T_o, \exists s \in T^*_{uo} \cup \emptyset, s.t \in SCS_N\}$
- Let N_i and N_j be two nodes of the Estimator, such that there is a directed arc $N_i \longrightarrow N_j$, labeled by a transition t_k (of T_o). N_i is a predecessor of N_j.
  - **Definition 1.** We call **Set of Entry Classes** of N_j (**SEC**(N_j)) the set of all classes of N_j which can be reached as the result of the firing of $t_k \in T_o$ starting from a class which belongs to a predecessor of N_j. The SEC of the initial node N_0 contains only the initial state class c_0.
  - **Definition 2.** The **Set of Shadow Classes** of the node N_j (**SSC**(N_j)) corresponds to the set of all classes in N_j reached following the firing of all achievable unobservable sequences starting from classes in **SEC**(N_j). In other terms, for each node N_j, we have: $SSC(N_j) = \mathcal{UR}(SEC(N_j))$.
  - **Definition 3.** Let N_i and N_j be two nodes of the Estimator.
    N_i and N_j are **equivalent** ($N_i \Leftrightarrow N_j$) iff $SEC(N_i) = SEC(N_j)$ and $SSC(N_i) = SSC(N_j)$.

5. The Estimator Building Algorithm

5.1. Algorithm

The Estimator building is done offline. The "skeleton" of the Estimator is obtained with performing an $\epsilon-$reduction on the class graph. As one goes along, the Estimator nodes are enriched with temporal behavioural constraints.
Here, N denotes the current node under elaboration, and NODES_TO_BE_TREATED the set of nodes which remain to be processed (list of successor nodes to be determined). In addition, TREATED_NODES is the set of nodes which were already treated.

1. /* *Initialization and establishment of the initial node* */
   {
   - NODES_TO_BE_TREATED $\longleftarrow \emptyset$, TREATED_NODES $\longleftarrow \emptyset$
   - create a node N_0
   - $SEC(N_0) \longleftarrow \{c_0\}$, $SSC(N_0) \longleftarrow \mathcal{UR}(c_0)$

   }
2. add N_0 to NODES_TO_BE_TREATED
3. /* *Treatment of a selected node, determination of successor nodes* */
   while (NODES_TO_BE_TREATED $\neq \emptyset$)
   {
   (a)
   - $N \longleftarrow$ pick a node from NODES_TO_BE_TREATED
   - NEXT$(N) \longleftarrow \emptyset$
   - add N to TREATED_NODES

   (b) compute $T_{N-obs-next}$ and $E_{N-obs-next} = evt(T_{N-obs-next})$

   (c) <u>repeat</u> for each class $c_i \in SEC(N)$:
   - determine the set SCS^{c_i} as well as the attributes (*duration interval* and *destination class*) of each sequence in SCS^{c_i}
   - add the elements of SCS^{c_i} to the candidate sequences table of c_i while ordering them in the increasing order of the maximum bounds of the duration intervals

   (d) <u>repeat</u> for each event $e_k \in E_{N-obs-next}$:
   {
   i. compute:
   $SEC_k = \{f(c_i, s.t_j);\ c_i \in SEC(N),\ t_j \in tr(e_k) \cap T_{N-obs-next}$ and $s \in T^*_{uo} \cup \epsilon$ such that $s.t_j \in SCS^{c_i}\}$

ii. */* if the set SEC_k does not correspond to any set of entry classes of an already established node, create a new node with SEC_k as set of entry classes or else add the node with SEC_k as set of entry classes to set of successor nodes of N */*

- repeat for all $N_i \in$ NODES_TO_BE_TREATED $\cup$ TREATED_NODES: if $SEC(N_i) = SEC_k$, then:
  - $N_k \leftarrow N_i$
  - goto 3(d)iii
- create a new node N_k
- $SEC(N_k) \longleftarrow SEC_k$, $SSC(N_k) \longleftarrow \mathcal{UR}(SEC_k)$

iii. add (N_k, e_k) to NEXT(N)

iv. repeat for each entry class $c_l \in SEC(N_k)$:

- repeat for each transition $t_j \in tr(e_k) \cap T_{N-obs-next}$:
  - add in the table of previous sequences of c_l all the sequences of $SCS_N^{t_j}$ which have c_l as *destination class*. With each of these sequences, associate its respective attributes (*duration interval* and *origin class*)

v. if $N_k \notin$ TREATED_NODES $\cup$ NODES_TO_BE_TREATED, then:

- add N_k to NODES_TO_BE_TREATED.

}

}

Proposition 1. *At any given time, the Estimator node containing the system's state can be determined with certainty.*

The proof of this proposition can be found in [6]. In this reference, the algorithm complexity matters are also dealt with.

5.2. Illustration on Our Case Study

The analysis of the TPN behavioral model of the system studied, with the TINA tool [2], shows that this PN is bounded and live. The class graph of our model is presented below (left part of figure 3). The Estimator is therefore acquired by applying the algorithm of section 5.1. (right part of figure 3). In this graph, dark disks represent entry classes and grey ones correspond to shadow classes.

The entry classes in the 4 nodes of our Estimator which are enriched with their previous and candidate sequences are shown in the tables below. N_0 has not previous nodes in the Estimator. Therefore, entry class c_0 in N_0 does not have any previous sequences.

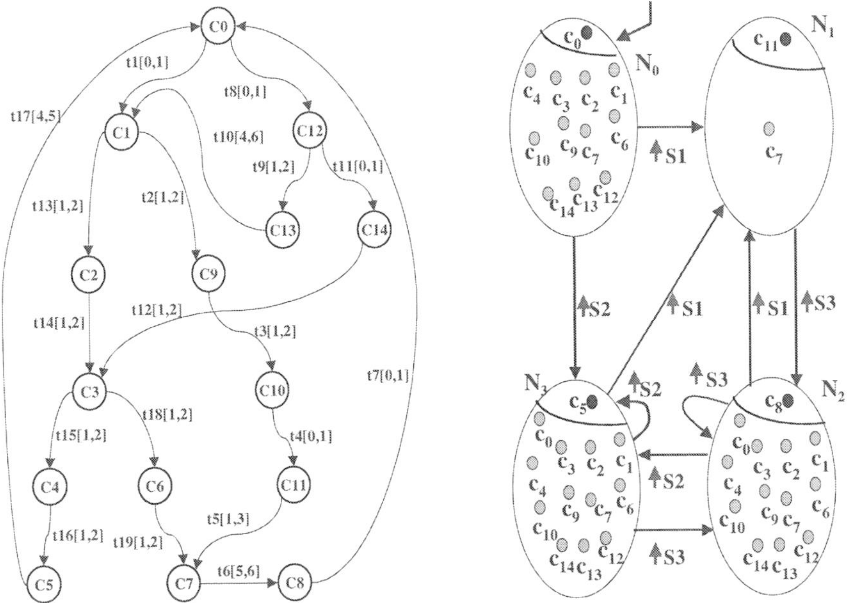

Figure 3. The State Class Graph and the Estimator of the System.

Table 2. Entry class C_0

Cand sequence	Dest	Max length
$t_1t_2t_3t_4$	c_{11}	7
$t_1t_{13}t_{14}t_{15}t_{16}$	c_5	10
$t_8t_9t_{10}t_2t_3t_4$	c_{11}	15
$t_1t_{13}t_{14}t_{18}t_{19}t_6$	c_8	17
$t_8t_9t_{10}t_{13}t_{14}t_{15}t_{16}$	c_5	18
$t_8t_{11}t_{12}t_{15}t_{16}$	c_5	18
$t_8t_9t_{10}t_{13}t_{14}t_{18}t_{19}t_6$	c_8	25
$t_8t_{11}t_{12}t_{18}t_{19}t_6$	c_8	25

6. Detectability and Diagnosability Analysis

In this section, discussed are some results about both the "Detectability" and the "Diagnosability" of a system for which a TPN behavioural model is provided as shown

Table 3. Entry class C_{11}

Prev sequence	Orig	Interval
$t_1t_2t_3t_4$	c_0	$[3, 7]$
$t_8t_9t_{10}t_2t_3t_4$	c_0	$[7, 15]$
$t_{17}t_1t_2t_3t_4$	c_5	$[7, 12]$
$t_{17}t_8t_9t_{10}t_2t_3t_4$	c_5	$[11, 20]$
$t_7t_1t_2t_3t_4$	c_8	$[4, 9]$
$t_7t_8t_9t_{10}t_2t_3t_4$	c_8	$[8, 17]$
Cand sequence	**Dest**	**Max length**
t_5t_6	c_8	5

previously in the chapter.

6.1. Detectability

6.1.1. Definition

A given system is said to be *detectable* w.r.t a set $\mathcal{F}$ of failures iff each failure in $\mathcal{F}$ can be *detected* upon a finite duration after its occurrence on the basis of the dated occurrences of observable events.

6.1.2. Properties

In this section, some necessary conditions for the *Detectability* property are proposed. $\mathcal{S}$ being a system and $\mathcal{F}$ a set of failures which can occur within $\mathcal{S}$.

- **Proposition 2.** $\mathcal{S}$ is detectable w.r.t $\mathcal{F}$ only if the following condition (denoted **(I)**) holds:
  $\forall\, c \in \mathcal{C},\ \forall\, e \in \Sigma_o,\ \nexists\, ((s_1, t_1), (s_2, t_2)) \in ((T_{un}^{*} \cup \emptyset) \times tr(e))$ such that:
  - $s_1 \neq s_2$
  - $f((c, s_1.t_1)) = f((c, s_2.t_2)) \neq \emptyset$
  - $int = [d_{min}(s_1.t_1), d_{max}(s_1.t_1)] \cap [d_{min}(s_2.t_2), d_{max}(s_2.t_2)] \neq \emptyset$
  - $P_F(s_1) = \emptyset$
  - $P_F(s_2) \neq \emptyset$

Table 4. Entry class C_5

Prev sequence	Orig	Interval
$t_1t_{13}t_{14}t_{15}t_{16}$	c_0	$[5,10]$
$t_8t_9t_{10}t_{13}t_{14}t_{15}t_{16}$	c_0	$[9,18]$
$t_{17}t_1t_{13}t_{14}t_{15}t_{16}$	c_5	[9,15]
$t_{17}t_8t_9t_{10}t_{13}t_{14}t_{15}t_{16}$	c_5	$[13,23]$
$t_7t_1t_{13}t_{14}t_{15}t_{16}$	c_8	$[6,12]$
$t_7t_8t_9t_{10}t_{13}t_{14}t_{15}t_{16}$	c_8	$[10,20]$
$t_8t_{11}t_{12}t_{15}t_{16}$	c_0	$[10,18]$
$t_{17}t_8t_{11}t_{12}t_{15}t_{16}$	c_5	$[14,23]$
$t_7t_8t_{11}t_{12}t_{15}t_{16}$	c_8	$[11,20]$
Cand sequence	**Dest**	**Max length**
$t_{17}t_1t_2t_3t_4$	c_{11}	12
$t_{17}t_1t_{13}t_{14}t_{15}t_{16}$	c_5	15
$t_{17}t_8t_9t_{10}t_2t_3t_4$	c_{11}	20
$t_{17}t_1t_{13}t_{14}t_{18}t_{19}t_6$	c_8	22
$t_{17}t_8t_9t_{10}t_{13}t_{14}t_{15}t_{16}$	c_5	23
$t_{17}t_8t_{11}t_{12}t_{15}t_{16}$	c_5	23
$t_{17}t_8t_9t_{10}t_{13}t_{14}t_{18}t_{19}t_6$	c_8	30
$t_{17}t_8t_{11}t_{12}t_{18}t_{19}t_6$	c_8	30

where $P_F(s)$ is the projection of s on F, ex. $P_{\{f_1,f_2\}}(a_1.a_5.f_2.b_2) = \{f_2\}$. The proof of this proposition is given in Appendix A..

- **Proposition 3.** $\mathcal{S}$ is detectable w.r.t $\mathcal{F}$ only if the following condition (denoted **(II)** holds:
  $\forall\, c \in \mathcal{C},\ \nexists(s_1, s_2) \in (T_{un}^* \cup \emptyset)$ such that:
  - $s_1 \neq s_2$
  - $f((c, s_1)) = f((c, s_2)) \neq \emptyset$
  - $int = [d_{min}(s_1), d_{max}(s_1)] \cap [d_{min}(s_2), d_{max}(s_2)] \neq \emptyset$
  - $P_F(s_1) = \emptyset$
  - $P_F(s_2) \neq \emptyset$

The proof of this proposition is given in Appendix B..

Table 5. Entry class C_8

Prev sequence	Orig	Interval
$t_1t_{13}t_{14}t_{18}t_{19}t_6$	c_0	$[11, 17]$
$t_8t_9t_{10}t_{13}t_{14}t_{18}t_{19}t_6$	c_0	$[15, 25]$
$t_{17}t_8t_9t_{10}t_{13}t_{14}t_{18}t_{19}t_6$	c_5	$[19, 30]$
$t_{17}t_1t_{13}t_{14}t_{18}t_{19}t_6$	c_5	$[15, 22]$
t_5t_6	c_{11}	$[3, 5]$
$t_7t_1t_{13}t_{14}t_{18}t_{19}t_6$	c_8	$[12, 19]$
$t_7t_8t_9t_{10}t_{13}t_{14}t_{18}t_{19}t_6$	c_8	$[16, 27]$
$t_8t_{11}t_{12}t_{18}t_{19}t_6$	c_0	$[16, 25]$
$t_{17}t_8t_{11}t_{12}t_{18}t_{19}t_6$	c_5	$[20, 30]$
$t_7t_8t_{11}t_{12}t_{18}t_{19}t_6$	c_8	$[17, 27]$
Cand sequence	**Dest**	**Max length**
$t_7t_1t_2t_3t_4$	c_{11}	8
$t_7t_1t_{13}t_{14}t_{15}t_{16}$	c_5	12
$t_7t_8t_9t_{10}t_2t_3t_4$	c_{11}	17
$t_7t_1t_{13}t_{14}t_{18}t_{19}t_6$	c_8	19
$t_7t_8t_9t_{10}t_{13}t_{14}t_{15}t_{16}$	c_5	20
$t_7t_8t_9t_{10}t_{13}t_{14}t_{18}t_{19}t_6$	c_8	20
$t_7t_8t_{11}t_{12}t_{15}t_{16}$	c_5	20
$t_7t_8t_{11}t_{12}t_{18}t_{19}t_6$	c_8	27

- **Proposition 4.** $\mathcal{S}$ is detectable w.r.t $\mathcal{F}$ only if the following condition (denoted **(III)**) holds:
  $\forall\, c \in \mathcal{C},\ \nexists(s_1, s_2) \in (T_{un}^* \cup \emptyset)$ verifying:

  - $s_1 \neq s_2$
  - $f((c, s_1)) = f((c, s_2)) = c$
  - $P_F(s_1) = \emptyset$
  - $P_F(s_2) \neq \emptyset$

  The proof of this proposition is given in Appendix C..

6.1.3. Diagnosability

6.1.4. Definition

When a given system $\mathcal{S}$ is detectable w.r.t a set $\mathcal{F}$ of failures, $\mathcal{S}$ is said to be *diagnosable* w.r.t $\mathcal{F}$ iff each failure in $\mathcal{F}$ can be *identified* upon a finite duration after its occurrence on the basis of the dated occurrences of observable events.

6.1.5. Properties

Some necessary conditions for the *Diagnosability* property are proposed. Here, $\mathcal{S}$ being a detectable system w.r.t to $\mathcal{F}$, a set of failures which may occur within $\mathcal{S}$.

- **Proposition 5.** $\mathcal{S}$ is diagnosable w.r.t $\mathcal{F}$ if the following condition (denoted IV) holds: $\forall\, c \in \mathcal{C},\ \forall\, e \in \Sigma_o,\ \nexists((s_1, t_1), (s_2, t_2)) \in (T^*_{un}) \times tr(e))$ such that:
  - $s_1 \neq s_2$
  - $f((c, s_1.t_1)) = f((c, s_2.t_2)) \neq \emptyset$
  - $int = [d_{min}(s_1.t_1), d_{max}(s_1.t_1)] \cap [d_{min}(s_2.t_2), d_{max}(s_2.t_2)] \neq \emptyset$
  - $P_F(s_1) \neq \emptyset$
  - $P_F(s_2) \neq \emptyset$
  - $P_F(s_1) \neq P_F(s_2)$

  The proof of this proposition is given in Appendix D..

- **Proposition 6.** $\mathcal{S}$ is diagnosable w.r.t $\mathcal{F}$ if the following condition (denoted V) holds:
  $\forall\, c \in \mathcal{C},\ \nexists(s_1, s_2) \in T^*_{un}$ verifying:
  - $s_1 \neq s_2$
  - $int = [d_{min}(s_1), d_{max}(s_1)] \cap [d_{min}(s_2), d_{max}(s_2)] \neq \emptyset$
  - $P_F(s_1) \neq \emptyset$
  - $P_F(s_2) \neq \emptyset$
  - $P_F(s_1) \neq P_F(s_2)$
  - $f((c, s_1)) = f((c, s_2)) \neq \emptyset$

  The proof of this proposition is given in Appendix E..

- **Proposition 7.** $\mathcal{S}$ is diagnosable w.r.t $\mathcal{F}$ if the following condition (denoted VI) holds:
  $\forall\, c \in \mathcal{C},\ \nexists(s_1, s_2) \in (T^*_{un})$ verifying:
  - $s_1 \neq s_2$

- $P_F(s_1) \neq \emptyset$
- $P_F(s_2) \neq \emptyset$
- $P_F(s_1) \neq P_F(s_2)$
- $f((c, s_1)) = f((c, s_2)) = c$

The proof of this proposition is given in Appendix F..

7. The Online Monitoring Process

7.1. Principle

The online monitoring process conjugates the observations gathered online with the static synthesis of the system's behaviour made within the Estimator. The monitoring is guided by the online observations (dated occurrences of observable events). Concretely, the elaborated mechanisms are of two types:

- passive mechanisms which intervene further to the occurrence of an observable event. They aim to determine the factual scenario which occurred through a discrimination process on the previous sequences of the entry classes in the attained node. This discrimination is done according to structural criteria (discerned event and origin class), then according to temporal criteria by correlating the event's occurrence date with the temporal data of the Estimator (occurrence intervals of previous sequences considered possible).
- anticipatory mechanisms which aim to "forecast" the monitoring process without waiting for the occurrence of some observable events. This is done by watching the maximum boundaries of the candidate sequences of the current node which are considered beforehand to be possible. This is, somehow, an adaptation of the watchdog technique.

The various mechanisms worked out allow the finding of the possible faults which may occur within the system (detection) and determine their types (diagnosis). These mechanisms are roughly defined in the algorithm presented hereafter. To better understand the various steps of the algorithm, the reader can refer to [6] and [7].

7.2. Online Monitoring Algorithm

When the system's state reaches a given node N_j of the Estimator, possibly from a node N_i following the occurrence of an observable event e corresponding to a transition $t \in T_o$ (the initial state of the system is in the node N_0, and more precisely, it can be represented by c_0), at a given instant, $\widetilde{SEC}(N_j)$ indicates the set of entry classes of N_j which could be attained. Similarly, $\widetilde{SSC}(N_j)$ is the set of shadow classes of N_j by which the system's state can pass through. Furthermore, e corresponds to the observable events which can occur, ϵ indicates the null event, N is the current node, PREV_NODE is the last node in which the system's state was, LAST_INDEX(c_j) is the index of the first candidate sequence, in the class c_j, of which the max duration was not exceeded and $\max_{c_j}\{\textit{maximal duration}\}$ is the maximal duration of the last candidate sequence in c_j.

1. /* *initialization* */
   - $N \longleftarrow N_0,\ e \longleftarrow \epsilon$
   - $\widetilde{SEC}(N) \longleftarrow SEC(N), \widetilde{SSC}(N) \longleftarrow SSC(N)$, PREV_NODE $\longleftarrow$ NULL

2. /* *processing of previous sequences* */
   if ($e \neq \epsilon$) /* *an observable event occurred* */
   {
   /* *initialization* */
   (a) $\widetilde{SEC}(N) \longleftarrow SEC(N)$ and $\widetilde{SSC}(N) \longleftarrow SSC(N)$
   (b) put all previous sequences of N in the state *possible*
   /* *checking of structural constraints* */
   (c) put all previous sequences other than those under form $s_i.t_j$ where $t_j \in tr(e)$, in the state *impossible*
   (d) put all previous sequences for which the *origin class* does not belong to $\widetilde{SEC}$(PREV_NODE), in the state *impossible*
   /* *checking of temporal constraints* */
   (e) for the remaining *possible* previous sequences, put in the state *impossible* those for which the relative time of occurrence of e does not belong to their respective duration intervals
   /* *searching of reachable classes* */
   (f) <u>repeat</u> for each class $c_i \in \widetilde{SEC}(N)$:
   <u>if (all</u> previous sequences of c_i are *impossible*)
      - eliminate c_i from $\widetilde{SEC}(N)$
      - $\widetilde{SSC}(N) \longleftarrow \mathcal{UR}(\widetilde{SEC}(N))$

   }

/* *(3) and (4) - Processing of candidate sequences* */

3. /* *initialization* */
   <u>repeat</u> for each class $c_j \in \widetilde{SEC}(N)$:
   - CURR_INDEX (c_j) $\longleftarrow 1$
   - put all candidate sequences of c_j in the state *possible*

4. /* *checking of the overtaking of max duration of candidate sequences* */
   while (no observable event has occurred)
   {

   repeat for each class $c_j \in \widetilde{SEC}(N)$:
   {
   (a) if (CURR_DATE $< \max_{c_j}\{max\ duration\}$), then do:
   {
   i. $k \longleftarrow$ CURR_INDEX (c_j)
   ii. if (CURR_DATE $>$ *max duration*(k^{th} candidate sequence))
      - put the kth candidate sequence in the state *impossible*
      - CURR_INDEX(c_j) $\longleftarrow$ CURR_INDEX(c_j) + 1

   }
   (b) else
   {
   - eliminate c_j from $\widetilde{SEC}(N)$
   - $\widetilde{SSC}(N) \longleftarrow \mathcal{UR}(\widetilde{SEC}(N))$

   }
   }

   }
5. /* *management of the transition further to an event detection* */
   as soon as an observable event e' occurs
   (a) PREV_NODE $\longleftarrow$ N
   (b) $N \longleftarrow f_N(N, e')$
   (c) $e \longleftarrow e'$
   (d) goto step (2)

7.3. Application

We propose now to supervise our system on the basis of the Estimator and by exploiting the information gathered online. Particularly, we shall supervise the occurrence of failures D_1 and D_2.
The train is initially on the segment T_0. As soon as the departure order is given, the train begins its cycle. The objective here is to track the system's state in time by supervising its behavior. For that, we will follow the monitoring algorithm developed in section 7.2.. Here, for lack of space, we shall limit ourselves to the presentation of a part of the online monitoring. Our goal being to give an idea about the general principle of the developed monitoring mechanisms.

The initial state of the system is in the node N_0 and can be represented precisely by c_0. The absolute date 7 is attained without any (observable) event being discerned. At that time, the maximum duration of $t_1t_2t_3t_4$ candidate sequence is attained. This sequence becomes, therefore, in the state *impossible*. $\uparrow S_1$ is discerned in the absolute date 9. The Estimator points out that the system's state switches over towards N_1 and more precisely towards its only entry class c_{11}. By analyzing the previous sequences of this class, one begins by putting in the state *impossible* sequence $t_1t_2t_3t_4$ since this sequence was an *impossible* candidate sequence in c_0 (as entry class of N_0). Then, all sequences which have not as origin class the class c_0 are put in the state *impossible*. Therefore only one previous sequence stays *possible* which is $t_8t_9t_{10}t_2t_3t_4$. Hence, one can detect that the factual scenario which occurred is: $A1b.A5b.A2b.A2d.A3d.\uparrow S_1$.
$\uparrow S_3$ is discerned at the absolute date 13. Thus, the system's state reaches the node N_2 and more precisely its only entry class c_8. By analyzing the previous sequences of this class, one begins by putting in the state *impossible* all sequences which do not have as origin class c_{11}. The only sequence which remains *possible* is t_5t_6 (discrimination between sequences using structural constraint ; as the origin class from which the system's state has evolved is c_{11}). We can deduce, therefore, that the factual scenario $A4b.\uparrow S_3$ occurred.
The absolute date 22 is attained without any event having taken place since the occurrence of $\uparrow S_3$. At that time, the maximum duration of the candidate sequence $t_7t_1t_2t_3t_4$ is attained ($9 = 22 - 13$). This sequence becomes, therefore, *impossible*.
Event $\uparrow S_2$ is caught at the absolute date 31. Then, the system's state switches towards N_3. It precisely reaches the entry class c_5. An analysis of the previous sequences of c_5 is done. First of all, all the sequences which don't have as origin class c_8 are eliminated. Two sequences remain *possible* which are $t_7t_1t_{13}t_{14}t_{15}t_{16}$ and $t_7t_8t_9t_{10}t_{13}t_{14}t_{15}t_{16}$. Both sequences end with a transition corresponding to the discerned observable event $\uparrow S_2$. Consequently, we cannot discriminate between them using this criterion. The temporal constraints are then checked: $\uparrow S_2$ is discerned at $31tu$ absolute date, i.e $9tu$ after the occurrence of $\uparrow S_3$. Only sequence $t_7t_1t_{13}t_{14}t_{15}t_{16}$ remains then *possible* since $9 \in [6, 12]$ and $9 \notin [10, 20]$. Hence, the factual scenario which is $A8d.A1d.A3b.A6b.A7b.\uparrow S_2$. Since $A3d$ corresponds to a failure (D_1), the associated alarm is triggered off.
Of course, according to the situation, it is not always possible to obtain a precise result for the detection/diagnosis process. In the conclusion, we list some further mechanisms which may potentially make more efficient the monitoring process when combined with the mechanisms discussed here.

8. Conclusion

A monitoring approach for discrete event timed systems is discussed in this chapter. The developed mechanisms are based on the timed signatures of the system's behaviour. They aim at tracking the system's state online while discerning the possible failures which may happen. We have shown the benefit of taking into account temporal aspects within the monitoring activity. The case study involved has enabled us to illustrate the approach on a reasonably-sized system. Moreover, this illustration shows the scalability's potential of the approach.

The monitoring process could be refined while taking into account the relative dates of the events' occurrences instead of absolute dates. Moreover, we intend to elaborate more deeply the backward analysis mechanisms (which "rewind" to more than one previous node) capable to resolve the uncertainty on the monitoring process. Before implementing our approach within a tool, some steps still have to be more formalized. Namely, transformation of TPN into timed automata [1] could offer a basis for using formal techniques [12], [4] to perform the diagnosis. In addition, the elaborated results as regards the diagnosability stand for necessary conditions only; the diagnosability analysis [11] has therefore to be pursued in order to obtain sufficient properties easily checkable. Finally, we intend to investigate the generalization of our approach to a distributed context [3]; therefore some new constraints have to be taken into account.

Appendix

To prove propositions 2-7, we will proceed by absurdity.

A. Proof of Proposition 2

Suppose that $\exists c \in \mathcal{C}$, $e \in \Sigma_o$, $(t_1, t_2) \in tr(e)^2$ and $(s_1, s_2) \in (T^*_{un} \cup \emptyset)$, $s_1 \neq s_2$ verifying:

- $f((c, s_1.t_1)) = f((c, s_2.t_2)) = c' \neq \emptyset$
- $int = [d_{min}(s_1.t_1), d_{max}(s_1.t_1)] \cap [d_{min}(s_2.t_2), d_{max}(s_2.t_2)] \neq \emptyset$
- $P_F(s_1) = \emptyset$
- $P_F(s_2) =\neq \emptyset$, let $f \in P_F(s_2)$

That can be depicted as in figure 4 below:

If the system's state reaches c, let's suppose that a scenario $\sigma.e$ ($\sigma \in \Sigma^*_{un} \cup \emptyset$) occurs, and let us suppose that the length of $\sigma.e$ is in the intersection int.
The only event discerned upon the occurrence of $\sigma.e$ is e after a relative duration in int.

Both sequences $s_1.t_1$ and $s_2.t_2$ have the same origin class (c), the same destination class (c'), are both in form $s.t$ ($s \in T_{un}$, $t \in tr(e)$), and scenario $\sigma.e$ fulfills the temporal constraints on both $s_1.t_1$ and $s_2.t_2$.

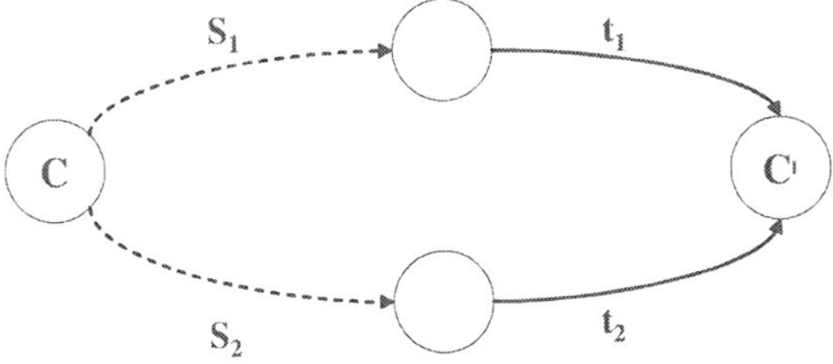

Figure 4. Counter-example of condition **(I)**.

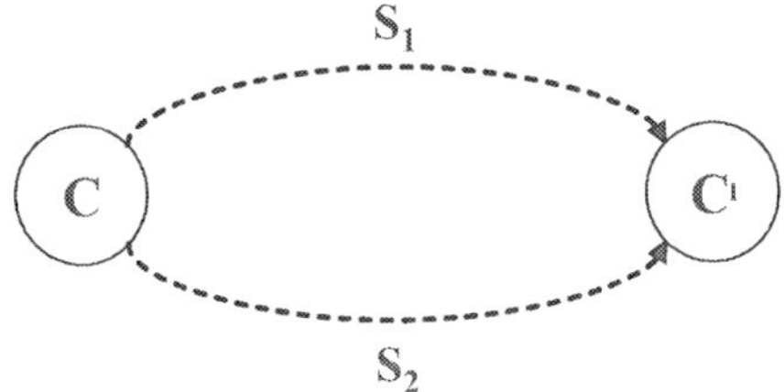

Figure 5. Counter-example of condition **(II)**.

Hence, there is no means which enables us to distinguish between these sequences, not even the forthcoming scenarios since $s_1.t_1$ and $s_2.t_2$ make the system's state reach the same destination class c'. Consequently, we cannot testify under any circumstances wether or not failure f has occurred during the passage of the system's state from c toward c'. That implies the non-detectability of $\mathcal{S}$ w.r.t $\mathcal{F}$.

B. Proof of Proposition 3

Suppose that condition **(II)** is not fulfilled i.e $\exists c \in \mathcal{C}$ and $(s_1, s_2) \in (T_{un}^* \cup \emptyset)$, $s_1 \neq s_2$ fulfills the conditions enumerated in proposition 2, as shown in figure 5.

If the system's state reaches c, we suppose that a scenario σ, relative to s_1 or to s_2 ($\sigma \in \Sigma_{un}^*$) occurs, during a duration in $[d_{min}(s_1), d_{max}(s_1)] \cap [d_{min}(s_2), d_{max}(s_2)] \neq \emptyset$ and that σ makes the system's state evolving from c toward c'. No event is thus observed and there is no means that enables us to identify the scenario which has actually occurred (that relative to s_1 or the one relative to s_2, or possibly another one), not even the forthcoming possible observable events. One cannot then realize whether some failures of $P_F(s_2)$ have happened which implies the non-detectability of $\mathcal{S}$ w.r.t $\mathcal{F}$.

C. Proof of Proposition 4

Suppose $\exists c \in \mathcal{C}$, and $(s_1, s_2) \in (T^*_{un} \cup \emptyset)$, $s_1 \neq s_2$ fulfilling the conditions enumerated in proposition 4 as depicted in figure 6.

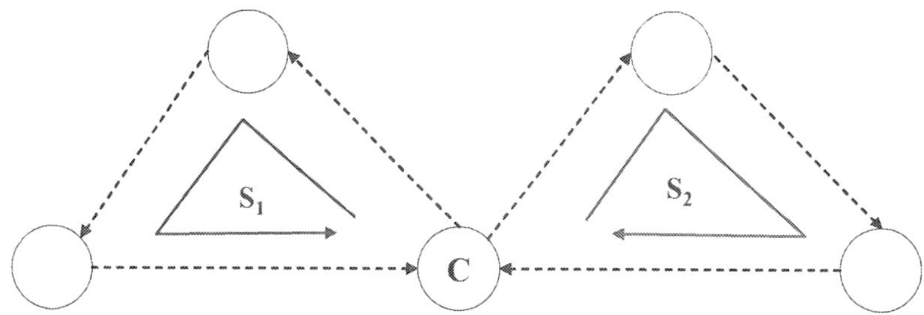

Figure 6. Counter-example of condition **(III)**.

According to the formal definition of a TPN, the transitions' enabling time are rational numbers ($\in \mathbb{Q}$), then the firing sequence durations are also rational numbers. It is therefore possible for:

- a dated firing sequence E_1, achievable starting from c, and having as firing support s_1, and
- another dated firing sequence E_2, achievable starting from c, and having as firing support s_2,

to find two integers n and p such that:

$$n * duration(E_1) = p * duration(E_2)$$

Indeed, if we consider a_1/b_1 ($a_1 \in \mathbb{N}$ and $b_1 \in \mathbb{N}^*$) the duration of E_1, and a_2/b_2 ($a_2 \in \mathbb{N}$ and $b_2 \in \mathbb{N}^*$) the duration of E_2, then we can take $n = b_1 * a_2$ and $p = b_2 * a_1$.

Now, let us assume that the system's state reaches c and that a scenario corresponding to E_1 occurs successively n times (or that a scenario corresponding to E_2 occurs successively p times). Consequently, the system's state moves from class c to regain it again. However, since s_1 and s_2 are composed of transition in T_{un} only, no sign allows us to identify whether the scenario which has occurred corresponds to $n * s_1$ ou $p * s_2$ (or possibly to another sequence in T_{un}). Identifying the actual scenario will not be

possible even with the upcoming evolution as both sequences $n * s_1$ and $p * s_2$ make the system's state reach c.
Hence, the detection process is not able to determine whether some failures (in $P_F(s_2)$) have occurred. In other terms, $\mathcal{S}$ is not detectable.

D. Proof of Proposition 5

Let us suppose that: $\exists c \in \mathcal{C}$, $e \in \Sigma_o$, $(t_1, t_2) \in tr(e)^2$ and $(s_1, s_2) \in (T^*_{un})$, $s_1 \neq s_2$ such that:

- $P_F(s_1) \neq \emptyset$
- $P_F(s_2) \neq \emptyset$
- $f((c, s_1.t_1)) = f((c, s_2.t_2)) \neq \emptyset$
- $P_F(s_1) \neq P_F(s_2)$
- $int = [d_{min}(s_1.t_1), d_{max}(s_1.t_1)] \cap [d_{min}(s_2.t_2), d_{max}(s_2.t_2)] \neq \emptyset$

When the system's state reaches c, imagine that a given scenario $\sigma.e$ ($\sigma \in \Sigma^*_{un}$) makes the system's state switch from c toward c'. Let us assume also that the duration of $\sigma.e$ is in the intersection $int = [d_{min}(s_1.t_1), d_{max}(s_1.t_1)] \cap [d_{min}(s_2.t_2), d_{max}(s_2.t_2)]$.
The only discerned event accordingly is e upon a relative delay in int.
Both Sequences $s_1.t_1$ and $s_2.t_2$ have as origin class c, as destination class c', and are both in the form $s.t$ ($s \in T_{un}$ and $t \in tr(e)$). Finally, the scenario $\sigma.e$ fulfills the temporal constraints on both sequences.
Then, there is no means which enables us to discriminate between these 2 sequences, not even the forthcoming scenarios since $s_1.t_1$ and $s_2.t_2$ make the system's state reach the same class c'. Thereby, the diagnosis process is not able to identify the type(s) of failures which may has (have) occurred. This implies that $\mathcal{S}$ is not dignosable w.r.t $\mathcal{F}$.
Nevertheless, if no other sequence in the form $s.t$ ($s \in T_{un}$) makes the system's state switch from c toward c', or if there is such a sequence but e occurs upon a delay outside the duration of $s_1.t_1$ and $s_2.t_2$, then upon the observation of e, one an conclude that some failures have surely occurred. Hence, the detection result is satisfying. Moreover, in terms of diagnosis (failures' identification), one can conclude that the set of failures which have occurred is equal to: $P_F(s_1)$ or $P_F(s_2)$.

E. Proof of Proposition 6

The proof can be made in the same way as for proposition 5. With the same hypothesis, the diagnosis process will not be able to conclude whether the scenario which has occurred is the one corresponding to s_1 or that which is relative to s_2 (or possibly another one).

F. Proof of Proposition 7

The proof can be made in the same way as for proposition 4. Except that here, the diagnosis process does not enable us to determine whether the failures which may have occurred are those in $P_F(s_1)$ or the ones from $P_F(s_2)$ (or possibly other failures).

References

[1] R. Alur and D. Dill, *A Theory of Timed Automata*, Theoretical Computer Science, Vol. 126, pp. 183-235, 1994.

[2] B. Berthomieu and P.O. Ribet and F. Vernadat, *The tool TINA – Construction of Abstract State Spaces for Petri Nets and Time Petri Nets*, International Journal of Production Research, 2004, 42, pp. 2741–2756.

[3] R.K. Boel and G. Jiroveanu, *Distributed contextual diagnosis for very large systems*, 17th Workshop on Discrete Event Systems (WODES'04), Reims, France, 2004.

[4] M. Bozga, C. Daws, O. Maler, A. Olivero, S. Tripakis, and S. Yovine, *Kronos: a model-checking tool for real-time systems*, 10th Conference on Computer-Aided Verification (CAV98), volume 1427 of LNCS, Springer, 1998.

[5] M. Diaz, *Les Réseaux de Petri, Modèles Fondamentaux*, Hermès, 2001.

[6] M. Ghazel, A. Toguyéni and P. Yim, *State Observer for DES under Partial Observation with Time Petri Nets, Journal of Discrete Event Dynamic Systems: Theory and Application*, Springer Ed., Vol. 19, Issue 2, pp. 137-165, 2009.

[7] M. Ghazel, *Monitoring of Discrete Event Systems with Time Petri Nets, PhD Thesis, LAGIS - École Centrale de Lille*, 2005.

[8] S. Lafortune and M. Sampath, *Discrete Event Systems Approach to Failure Diagnosis: Theory and Applications*, 11th International Workshop on Principles of Diagnosis, 2000.

[9] P. Merlin, *A Study of the Recoverability of Computer System*, PhD thesis, university of California, 1974.

[10] D.N. Pandalai and L.E. Holloway, *Template language for fault monitoring of discrete event processes*, IEEE Transactions on Automatic Control, May 2000.

[11] M. Sampath, R. Sengupta, S. Lafortune, K. Sinnamohideen, and D. Teneketzis, *Diagnosability of discrete event systems, IEEE Transactions on Automatic Control, Vol. 40(9)*, September 1995.

[12] S. Tripakis, *The formal analysis of timed systems in practice, PhD thesis*, Université Joseph Fourrier de Grenoble, 1998.

[13] S. Tripakis, *Fault diagnosis for timed automata*, Formal techniques in real time and fault tolerant systems, LNCS2469, Springer, 2002.

[14] T. Ushio and I. Onishi and K. Okuda, *Fault Detection Based on Petri Net Models with Faulty Behaviors*, Proceeding of IEEE International Conference on Systems, Man, and Cybernetics, 1998, pp. 113–118.

[15] S.H. Zad and R.H. Kwong and W.M. Wonham, *Fault Diagnosis in Timed Discrete-Event Systems*, 38th conference on Decision & Control, Phoenix, Arizona - USA, pp. 1756–1761, 1999.

In: Fault Detection: Theory, Methods and Systems ISBN: 978-1-61728-291-1
Editor: Léa M. Simon, pp. 97-176

Chapter 3

FAULT DETECTION AND DIAGNOSIS WITH STATISTICAL AND SOFT COMPUTING METHODS

Juan Pablo Nieto González * ***and Luis Eduardo Garza Castañón*** †
Monterrey Institute of Technology, Mexico

Abstract

Most of the research work that has been done on the field of fault detection and diagnosis has used the model based approach. This chapter presents an alternative way of carrying out a complete fault detection and diagnosis system using statistical and soft computing methods. This proposal is based only on the system's or process' history data treatment. The motivation of using process history data is basically: 1. To obtain an approach that could take into account variables correlations, that at first sight could not be so clear even for expert designers, 2. To develop a diagnostic system that learns normal operation mode directly from the process, instead of having a model based diagnostic system which depends on the expertise of the designer to manage the complexity of the system when modeling. The advantage of having a process history data based approach over a model based framework is the relatively easy way to obtain data from automated industrial processes. In most of the modern systems, can be very difficult to obtain an exact model due to the big quantity of information needed and the variables correlations, which can cause false alarms,

*E-mail address: juan.pablo.nieto@itesm.mx
†E-mail address: legarza@itesm.mx

indicating a wrong faulty component or system. Nevertheless, this kind of approaches combining statistical and soft computing methods can be supported or complemented with model based methods, in order to have a more powerful diagnosis method combining the expertise and mastery of the designer and those hidden behaviors that many systems exhibit. In this chapter it is shown how statistical methods applied in a straightforward way and combined with soft computing methods such as artificial neural networks and fuzzy logic, are ideal tools for doing diagnosis. This translates to finding the root cause of the problem using only process history data as a prior knowledge of a system, no matter if it is linear or nonlinear. This knowledge is used to give a final diagnosis, in complex scenarios whith noise presence and correlated variables, that could easily mislead to false alarms or wrong diagnosis. The organization of this chapter is as follows. First of all it is presented an introduction of why a fault diagnosis is necessary. Then a classification of fault diagnosis methods is given. After that, a presentation of the mathematical tools used is shown and then how they have been tailored in our research, in order to build complete fault detection and diagnosis systems for several applications. We present case studies that show promising results using the algorithms proposed. Finally the conclusion over this chapter is given.

Keywords: Fault Detection, Diagnosis, Autoassociative Neural Network, ANFIS, non-linear PCA, Correlation, Softcomputing, Vehicle Diagnosis, Power Systems Diagnosis.

1. Introduction

1.1. Why a Fault Diagnosis Is Necessary

Due to the building and utilization of machines since the industrial revolution, engineers have been concerned about their conditions to work properly. In the beginning of the utilization of machines the only way to know and learn about malfunctions of them, was the biological senses. They used to learn using for instance their hands to feel over heat or vibrations as well as smelling to detect fumes. Sometime later it was introduced the concept of the measuring devices which provided more exact information about important physical variables. In our days this concept of measuring devices still is in use in the same way with the application of sensors. The use of sensors provides the benefit of measuring a physical variable, but as they are also systems of measurements are exposed to damages that can be confused with process alarms. In addition to that, the potential for faults in sensors become more critical when they are applied in the automatic control of the machines, where the effects of such malfunctions may be observed directly as a loss of production either by bad quality or a machine brake down, or in a devastating result as an accident.

As a result of the research and the needs of treatment of data which lead the industry to reach better quality in the final products, and the automation of several processes, began the use of computers and microprocessors. Then, as the use of computers and microprocessors were expanded to the automation and monitoring of several processes, it was required a method of supervising many control loops. In our days almost everywhere we can find machines that help humans to do their works. For instance the automatic pilot of an airplane and the automation of all its complex subsystems, the computer on board of a car that tells the location and the measuring of certain variables as air pressure of the tires, the levels of all the fluids or when it tells that one or some components of the car needs to be verified by the technician. But in the same way as those systems need to be diagnosed to take some decisions, it is necessary to have some kind of supervision that tells human the behavior of certain or several physical variables to know if a system is working properly or if there is a fault present on it. And more yet, tells human where the fault is, that means which component or components are not working properly and why.

Some definitions commonly used in fault detection and diagnosis field are:

- Fault: A deviation not allowed of at least one characteristic, property or parameter of a system from an acceptable, usual or standard condition.
- Failure: A permanent interruption of the ability of a system to carry out a required function under specific operation conditions.
- Fault Detection: It is the determination of fault present in a system and its detection time.
- Fault Diagnosis: It is the determination of the kind, size, location and time of a fault.

1.2. Fault Diagnosis Methods

As processes become more complex, the monitoring of them is very important in order to improve process performance, efficiency and product qual-

ity. Monitoring of industrial processes plays a substantial role in system safety, availability and production quality. Early detection of faults can help to avoid major breakdowns and incidents. In order to tackle those problems, fault detection and diagnosis have been lately very active research domains.
There exist many research works related with fault detection which are based in analytical methods, artificial intelligence and statistical methods. [1] classifies fault detection and isolation methods in three groups (see Fig.1): 1) Quantitative Model Based, 2) Qualitative Model Based and 3) Process History Based.

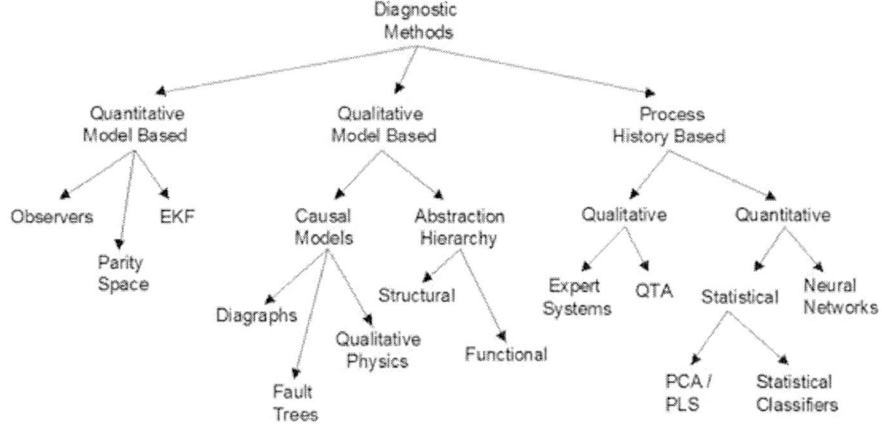

Figure 1. Classification of Diagnostic Methods.

Quantitative Model Based fault detection methods are based on a mathematical model of the system. The occurrence of a fault is captured by discrepancies between the observed behavior and the behavior that is predicted by the model. These approaches make use of the state estimation, parameter identification techniques, and parity relations to generate residuals. Fault localization then rest on interlining the groups of components that are involved in each of the detected discrepancies. However, it is often difficult and time-consuming to develop accurate mathematical models that characterize all the physical phenomena occurring in industrial processes.
Qualitative Model Based fault detection methods use symbolic reasoning which generally combines one of three possible kinds of knowledge with graph theory, to analyze the variables relationships of a system. An advantage of these methods is that an explicit model of the system to be diagnosed is not necessary.

Knowledge-based approaches such as expert systems may be considered as alternative or complementary approaches where analytical models are not available.

Process History Based fault detection methods only require a big quantity of historical data process. There exist several ways in which these data can be transformed and presented as prior knowledge of a system. These transformations are known as feature extraction and can be qualitative as that used by expert systems and qualitative trend analysis methods or quantitative as that used in neural networks, PCA, PLS or statistical pattern recognition.

In general there are a many techniques and combinations of them used in fault detection and diagnosis, but it will depend on the problem 's nature to select the technique that has the best performance. An important number of characteristics called fault sensitivity are related to the detection technique behavior and they are very important in the selection of the technique used.

Such characteristic among others are:

- Ability to detect reasonably small size faults
- Detection speed
- Ability to operate with noise presence
- Ability to operate with errors in model system
- Efficiency to obtain none or the minimum of false alarms

Fault detection systems can be classified in two big categories according to their operation in plants as: 1) Fault detection systems online and 2) Fault detection systems offline, and both perform in general the following tasks:

1. Fault Detection: It is an indicative that something wrong is happening in a system being monitored.

2. Fault Isolation: It is the determination of the exact location of a fault (it gives the exact faulty component).

3. Fault Identification: It is the determination of the fault magnitude.

The isolation and identification tasks combined together are known as fault diagnosis. It is important to remark that these three tasks are hard due to:

- Complex faults scenarios where mechanical, electrical and electronic devices are interrelated and working together with PCs or PLCs.
- Uncertainty in information
- Lack of domain knowledge

2. State of the Art

Several papers present evidence in which fault detection and diagnosis in different processes are carried out using either, an individual technique or making a combination of different techniques, taking advantage of the best characteristics of each to achieve a better general behavior of the fault detection process. For instance [2] presents a comparative study in the monitoring of hybrid systems, where the continuous part is modeled by Bond Graph and the discrete part is modeled by Petri Nets. [3] proposes a structure of a hybrid fault diagnosis system which integrates Signed Directed Graph, Artificial Neural Network and dynamic simulation. [4] introduces a hybrid method based on hybrid system theory, which combines knowledge based methods and model base methods. [5] proposes a process monitoring which is composed of three parts, pre-analysis, visualization and diagnosis, where the proposed method integrates PCA, FDA and clustering analysis taking advantage of each technique for a complete solution. [6] describes plant devices, sensors, actuators and diagnostic tests as stochastic finite state machines, by assigning transition probabilities and marginal probabilities to safe and fault events, through simple composition rules it is possible to determine the feasible configuration of alarms and their conditional probability given any event. [7] combines the use of signed directed graph to make a classification model, PCA and fuzzy knowledge to form a qualitative and quantitative model and compares the grade of the patterns needed to be diagnosed to the given fault patterns. [8] proposes a method based on the interaction between AI and control techniques. It uses a causal graph representation of the process enabling decomposition into subsystems, reducing the diagnostic computational complexity. After that, at local level, FDI techniques based on numerical residual generation and analysis are carried out. [9] presents a useful method when measures on the input signals cannot be done due an inexistent sensor to measure a specific input or because it is impossible to do measurements due to the nature of the system itself. The diagnosis system takes plant output signals and combines its variances and then uses a discriminant analysis upon the resultant features to carry out the diagnosis. In [10] PCA and statistical control chart are used to detect process operating faults on an industrial rolling mill reheating furnace. The Q statistic and Hotelling T^2 statistic are used to calculate the control limits of the statistical control chart. [11] proposes a fault diagnosis model which extracts multi-dimension features from the detected signal to supervise the different features of the signal simultaneously. [12] describes a fault parametrization based on hybrid automata models and considers both abrupt failures and gradual degradation of system components. The approach also addresses the computational problem of coping with a large amount of sensor data by using a discrete event model of the system so as to focus distributed signal analysis on when and where to look for signatures of interest.

The previously mentioned methods are powerful if the designer considers all the variables of the modeled-process being represented. However, it is very difficult to establish all the model equations as the system grows in complexity, as in electrical power systems for instance. In this domain, the need to develop more powerful approaches has been recognized, and hybrid techniques that combine several reasoning methods start to be used [13]. This approach incorporate model based diagnosis and signal analysis with neural networks. [14] uses Bayesian networks (BNs) to estimate the faulty section of a transmission power system. Simplified models of BNs with Noisy-Or and Noisy-And nodes are proposed to test if any transmission line, transformer, or busbar within a blackout area is faulty. In [15] an investigation is performed about the use of logistic regression and neural networks to classify fault causes. This paper also discusses about data insufficiency, imbalanced data constitution and threshold setting. Ren and Mi [16] propose a procedure for power systems fault diagnosis and identification based on Petri Nets and coding theory. They tested the approach with simulations over the IEEE 118-bus power system and highlight the great advantage to handle very easily future expansions. In [17] a Fault diagnosis system is presented, based on multi-agent systems. By using a negotiation mechanism between decision-making agent and a cooperative agent, fault diagnosis results can be obtained. [18] proposed an approach based on four independent artificial neural networks (ANN) for real time fault detection and classification in power transmission lines. The technique uses consecutive magnitude current and voltage data at one terminal as inputs to the corresponding ANN. The ANN outputs are used to indicate simultaneously the presence and the type of the fault. [19] developed a methodology using wavelet transform for phase to ground fault detection in primary distribution systems, but it is an efficient methodology only for single phase fault detection in unbalanced distribution systems.

Another challenging domain for fault diagnosis is a vehicle. A vehicle is a complex engineering system where there are a large number of sensors, controllers, and computer modules embedded in the vehicle that collect abundant signals. The area of vehicle fault diagnosis relies on the processing of such signals that many times include non linear and noisy characteristics. For this reason, a threshold has to be used to declare the existence of a fault. However the computation of thresholds has to be done carefully to perform the rejection of noisy signals and allow the detection of real faults. In the past, different combinations of techniques have been done to deal with this problem. For instance [20] uses a bank of observers combined with estimators, then the residuals are compared using a sequential test of Wald, but it is pointed out that this method has limitations in presence of high measurement noise, but this test needs to be done every instant and additionally it needs the definition of two thresholds in order to obtain a low false alarms rate and a high rate of detection.

[21] applies a dynamic Bayesian network on the residuals generated by multiple simple models of a vehicle, but propagating the dynamic Bayesian network requires computing and storing an exponentially large probability distribution giving thus good results only for small number of faults. [22] is based on [21], a Bayesian network to process model residuals. The dynamic Bayesian network models the temporal behavior of the faults and determine fault probabilities taking into account the measurement noise presence on the system. But the great disadvantage is that it needs the joint probabilities of all possible fault combinations. It takes into account noise presence, but with the assumption that the Bayesian network must sample the residuals slower than the noise dynamics. Thus, combining the techniques mentioned above the performance of a diagnosis process could be improved and make a fault tolerant control as shown in [23]. [24] uses a sliding mode observer bank and a bicycle vehicle model using small angle approximations to represent the vehicle behavior. [25] uses a model based method in combination with fuzzy logics to detect, isolate and determine the fault type. Three estimates of the vehicle system behavior are generated and compared to the actual measurements generating four residuals to be monitored each with a set of fuzzy rules interpreting these outputs as the probability of the occurrence of a specific fault. [26] is based on residual gen-

eration using eigenstructure assignment to make detection insensitive to model uncertainties. [27] introduces a fault tolerant sensor system using a Kalman Filter and a two track model designed to create analytical signals replacing the faulty sensor.

3. Mathematical Preliminaries

Due to evidence shown in the last section, it has been thought an alternative solution to model the behavior of the plant. The quantitative methods mentioned above are powerful when the designer considers all the variables in the process being model. So it is clear that it is very difficult to establish all the model equations as system complexity grows and another important point is that all those methods do not take into account correlation between variables. That is why process history methods are getting more importance in the fault detection field. They have some techniques that are based on reasoning methods or probabilistic and statistic methods, which do not need a complete mathematical model of the whole environment and could work with the presence of noise and variables correlation. The reason is that probability and statistics provides a way of summarizing the uncertainty that comes from our laziness and ignorance at the moment of modeling a plant. An advantage over model based methods is that the frameworks based on process history data only needs a good historical data set of normal system operation, which in practice it is easy to obtain for computer controlled industrial processes. The methods shown in this chapter are easy to implement and to adapt because when original process changes, it is only required to modify the original data base instead of develop a new mathematical model from it. Another advantage is that the use of statistical methods like PCA allows analysis with less quantity of data, but keeping the original correlation between variables. This chapter integrates some of our research results on applying Process History Methods in Fault Detection and Diagnosis. Thus, in the following we present a brief description of all the mathematical tools that we have used alone or combined in order to build a complete diagnosis system for different applications.

3.1. Principal Components Analysis (PCA)

The principal component analysis is concerned with explaining the variance and covariance structure of a set of variables, through a few linear combinations of these variables. The general objectives are basically: data reduction and interpretation.

PCA decomposes the X original data matrix with dimension $m \times n$ (m number of samples and n number of variables) as:

$$X = t_1 p_1^T + t_2 p_2^T + \cdots + t_n p_n^T + E = T_n P_n^T + E \tag{1}$$

t_i vectors are called the scores of the principal components and have information on how the samples are related to each other. p_i are the eigenvectors of the covariance matrix of X, and are known as the loads of the principal components. They have information on how the variables are related to each other. In fact principal components analysis splits X matrix in two parts, one describes the system variation and the other captures noise or information not modeled. The X matrix could often be estimated using only A ($\leq n$) principal components instead of n variables as

$$\hat{X} = \sum_{i=1}^{A} t_i p_i^T + e \tag{2}$$

Where e is the residual. PCA is scale dependent, thus when the variables are measured in different scales or on a common scale with different ranges, they are often standardized. Another important issue is the minimum quantity of components needed to explain the data. The number of PC to retain in order to represent the maximum variance depends on the data and the existing correlation between the variables such that there are several decision criteria. [28] proposes to consider the amount of total sample variance explained, the relative sizes of eigenvalues or the use of scree plots. Thus the number of principal components should be equal or less than the variables of X. When the maximum variance of data is explained with the first two principal components, samples are on a plane and a constant density ellipse could be formed by them.

Figure 2 shows the plane and constant density ellipse formed by two principal components, where the first principal component is the one that has the major data variation, while the second one is the next with the major data variation of the rest and is orthogonal to the first one. Thus, PCA model is able to describe significant variations in fewer dimensions than the original n variables does.

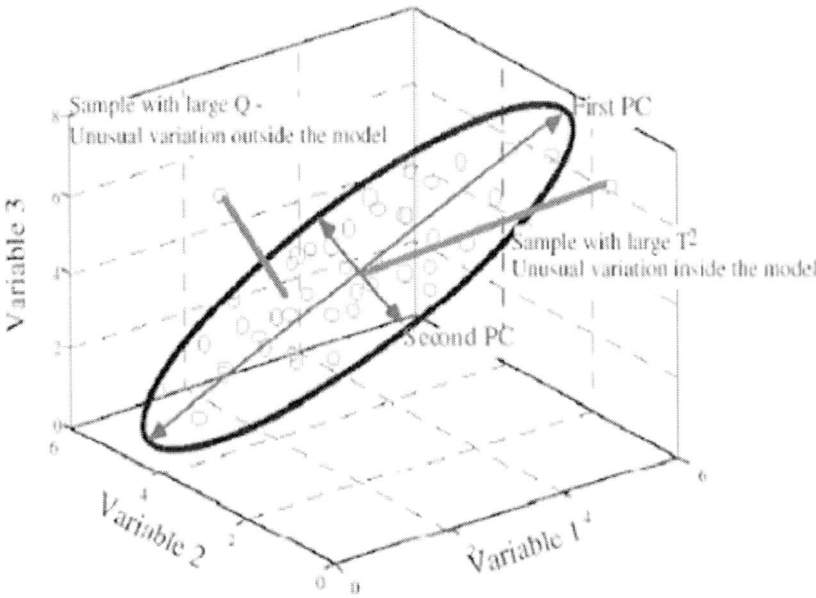

Figure 2. Plane and constant density ellipse formed by two principal components.

3.2. Control and Contribution Charts in Statistical Process Control

Generally, there are two statistics to define the action and warnings limits used in multivariate control charts. The first statistic is Hotelling's T^2 as follows:

$$T_i^2 = X_i P \Lambda P^T X_i^T \sim \frac{A(n-1)}{n-A} F_{A,n-A} \tag{3}$$

Where X_i is the vector containing the data matrix X at sample time i, and Λ is a diagonal matrix containing the inverse of the eigenvalues of the PC scores. T^2 is a statistical measure of the multivariate distance of each observation from the center of the data set. This is an analytical way to find the most extreme points in the data. Thus, an out of control signal is identified if

$$T_i^2 > \frac{A(n-1)}{(n-A)} F_{A,n-A,\alpha} \tag{4}$$

confidence limit α typically takes the value of 0.05 or 0.01 for the limits. The second metric used in process monitoring to identify non-conforming operation is the Q statistic (also referred as Squared Prediction Error, SPE). The Q statistic is defined to be the quadratic form of the residuals, that is the squared difference between the observed values and predicted values from the nominal or reference models:

$$Q_i = e_i e_i^T = \sum_{j=1}^{k} (x_{ij} - \hat{x}_{ij})^2 \tag{5}$$

And its upper limit (UL) is given by a chi-square distribution with $p-A$ degrees of freedom

$$UL = \chi^2_{p-A}(\alpha) \tag{6}$$

Q is the statistic that measures lack of fit of a model to data. Under the assumption that the linear PCA is valid, the Q statistic defines the Euclidean distance of the position of an observation from the hyperplane formed by the PCA model. See figure 2. In addition we could determine which variable of the process being analyzed is responsible for the unusual Q behavior, looking at a chart showing the contribution of each input to the Q statistic. This chart is known as the contribution chart and includes all process variables and their corresponding PCA scores in its axis.

3.3. Automatic Statistical Limits Computation

[11] gives an algorithm to extract the statistical boundary vectors of a multi-dimensional feature extraction. In our application, we have done a modification of that algorithm. Instead of doing multi-dimension feature extraction, here we work just with the statistical mean of the system variable being measured. Thus, the algorithm of inductive learning is used to obtain the statistical boundary vectors of $w_{max}(t_1,t_n,i)$ and $w_{min}(t_1,t_n,i)$ from a matrix which m_i rows are the different subsystems forming an entire process and the n columns present the changes of the statistical mean of the subsystem variable being measured as time changes from time t_1 to time t_n. The algorithm is shown below.

1. Initialize the statistical boundary vectors $w_{max}(t_1,t_n,i) = [w(t_1,t_2,1) \cdots w(t_1,t_2,m_i)]^T$, $w_{min}(t_1,t_n,i) = [w(t_1,t_2,1) \cdots w(t_1,t_2,m_i)]^T$ and the counter $j=0$.

2. Calculate $w_{max}(i) = [max(w_{max}(t_1,t_n,1), w_{max}(t_n,t_{n+1},1) \cdots max (w_{max}(t_1,t_n,m_i), w_{max}(t_n,t_{n+1},m_i)))]^T$ and calculate $w_{min}(i) = [min (w_{min}(t_1,t_n,1), w_{min}(t_n,t_{n+1},1) \cdots min(w_{min}(t_1,t_n,m_i), w_{min}(t_n, t_{n+1},m_i)))]^T$. If $w_{max}(i) = w_{max}(t_i,t_n,i)$ and $w_{min}(i) = w_{min}(t_i,t_n,i)$, then $j = j+1$, else $j = 0$.

3. If $j \geq V_1$, go to step(4), else $w_{max}(t_1,t_n,i) = w_{max}(i)$, $t_n = t_{n+1}$, go to step (2)

4. Output $w_{max}(t_1,t_n,i)$, $w_{min}(t_1,t_n,i)$, t_n, exit.

Where $w_{max}(t_1,t_n,i)$ and $w_{min}(t_1,t_n,i)$ in our research are used as the desired statistical limits for the statistical mean of the system variable being measured.

3.4. Discriminant Analysis (DA)

Discrimination and classification are multivariate techniques concerned with separating distinct sets of objects (or observations) and with allocating new objects (observations) to previously defined groups. Thus, the immediate goals of discrimination and classification, respectively are as follows:

1. To describe, either graphically or algebraically, the differential features of observations from several known collections. It tries to find whose numerical values are such that the collections are separated as much as possible

2. To sort observations into two or more labeled classes. The emphasis is on deriving a rule that can be used to optimally assign new observations to the labeled classes.

It is convenient to label the classes π_1 and π_2. The observations are ordinarily separated or classified on the basis of measurements on, for instance, p associated random variables $X' = [X_1, X_2, ..., X_p]$. The observed values of X differ to some extent from one class to the other. Classification rules are usually developed from learning samples.

3.4.1. Fisher Discriminant Function - Separation of Populations

Fisher's discriminant function idea is to transform the multivariate observations x to univariate observations y such that the $y's$ derived from populations

π_1 and π_2 are separated as much as possible. Fisher's discriminant function suggests to take linear combinations of x to create $y's$ because they are simple enough functions of the x to be handled easily. Fisher's approach does not assume that the populations are normal. It does, however, implicitly assume that the population covariance matrices are equal, because a pooled estimate of the common covariance matrix is used. A fixed linear combination of the x's takes the values $y_{11}, y_{12}, ..., y_{1n_1}$ for the observations from the first population and the values $y_{21}, y_{22}, ..., y_{2n_2}$ for the observations for the second population.

Thus an allocation rule based on fisher's discriminant function is:

Allocate x_0 to π_1 if

$$\hat{y}_0 = (\bar{x_1} - \bar{x_2})' S^{-1}_{pooled} x_0 \geq \hat{m} \tag{7}$$

where

$$\hat{m} = \frac{1}{2}(\bar{x_1} - \bar{x_2})' S_{pooled} (\bar{x_1} + \bar{x_2}) \tag{8}$$

or rearranging 7 and 8

$$\hat{y}_0 - \hat{m} \geq 0 \tag{9}$$

Allocate x_0 to π_2 if

$$\hat{y}_0 < \hat{m} \tag{10}$$

or

$$\hat{y}_0 - \hat{m} < 0 \tag{11}$$

Fisher's linear discriminant function in 7 was developed under the assumption that the two populations, whatever their form, have a common covariance matrix.

3.5. Probabilistic Neural Network

Probabilistic Neural Network (PNN) are conceptually similar to K-Nearest Neighbor (KNN) models [28], [29]. The basic idea is that a predicted value of an item is likely to be about the same as other items that have close values of the predictor variables.

From figure 3 it is assumed that each case in the training set has two predictor variables, x and y. The cases are plotted using their x,y coordinates as shown in the figure. Also we assume that the target variable has two categories, positive which is denoted by a square and negative which is denoted by a dash. It can be noted that the triangle is positioned almost exactly on top of a dash representing

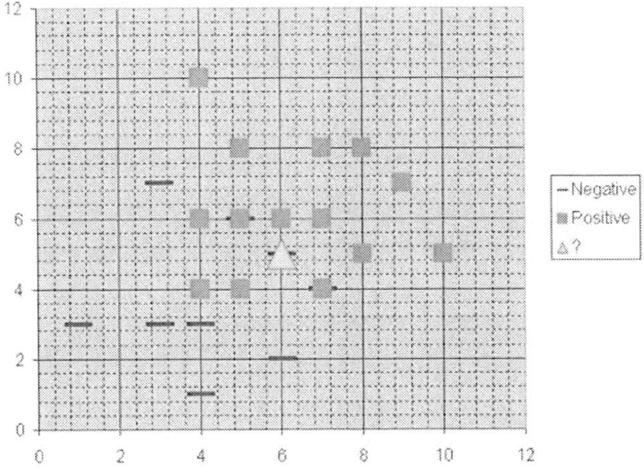

Figure 3. PNN are conceptually similar to KNN.

a negative value. But that dash is in a fairly unusual position compared to the other dashes which are clustered below the squares and left of center. So it could be that the underlying negative value is an odd case. The nearest neighbor classification will depend on how many neighboring points are considered. If 1-NN is used and only the closest point is considered, then the new point should be classified as negative since it is on top of a known negative point. On the other hand, if 9-NN classification is used, the closest 9 points are considered and then the effect of the surrounding 8 positive points may overbalance the close negative point. A probabilistic neural network builds on this foundation and generalizes it to consider all of the other points. The distance is computed from the point being evaluated to each of the other points, and a radial basis function (RBF) (also called a kernel function) is applied to the distance to compute the weight (influence) for each point. The radial basis function is so named because the radius distance is the argument to the function. $Weight = RBF(distance)$ the further some other point is from the new point, the less influence it has. Different types of radial basis functions could be used, but the most common is the Gaussian function. The PNN architecture is shown in Figure 4. The model has two layers: radial basis layer and competitive layer.

There are Q input vector/target vector pairs. Each target vector has K elements. One of these elements is 1 and the rest is 0. Thus, each input vector is

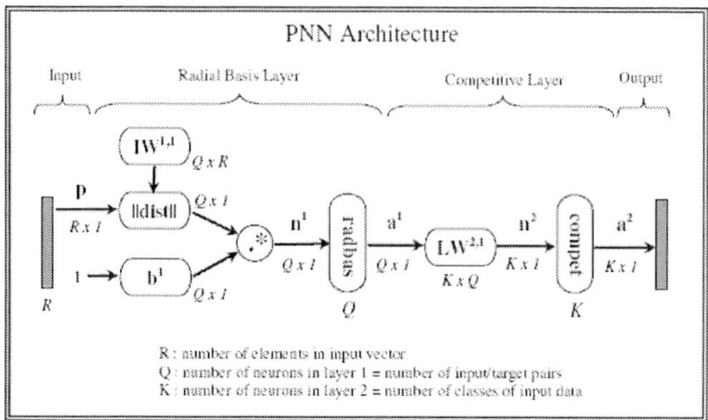

Figure 4. PNN architecture.

associated with one of K classes. When an input is presented the $||dist||$ box produces a vector whose elements indicate how close the input is to the vectors of the training set. An input vector close to a training vector is represented by a number close to 1 in the output vector a^1. If an input is close to several training vectors of a single class, it is represented by several elements of a^1 that are close to 1. Each vector has a 1 only in the row associated with that particular class of input, and 0's elsewhere. The multiplication Ta^1 sums the elements of a^1 due to each of the K input classes. Finally, the second layer, produces a 1 corresponding to the largest element of n^2, and 0's elsewhere. Thus, the network has classified the input vector into a specific one of K classes because that class had the maximum probability of being correct.

3.6. Correlation Matrix and Eigenvalues Definitions

Correlation Matrix Definiton. A Correlation matrix describes correlation among M variables. It is a square symmetrical $M \times M$ matrix with the (ik)th element equal to the correlation coefficient r_{ik} between the (i)th and the (k)th variable. The correlation coefficient is obtained as

$$r_{ik} = \frac{\sum_{j=1}^{n}(x_{ji} - \bar{x_1})(x_{jk} - \bar{x_k})}{\sqrt{\sum_{j=1}^{n}(x_{ji} - \bar{x_i})^2}\sqrt{\sum_{j=1}^{n}(x_{jk} - \bar{x_k})^2}} \tag{12}$$

The diagonal elements (correlations of variables with themselves) are always equal to 1.00 [30].

Eigenvalue Definition. Let A be a $k \times k$ square matrix and I be the $k \times k$ identity matrix. Then the scalars

$$\lambda_1, \lambda_2, \ldots, \lambda_k \tag{13}$$

satisfying the polynomial equation

$$|A - \lambda I| = 0 \tag{14}$$

are called the eigenvalues or characteristic roots of a matrix A. The equation $|A - \lambda I| = 0$ is called the characteristic equation, thus similar matrices and A and its transpose matrix have same eigenvalues [30].

3.7. Multidimensional Scaling

Multidimensional Scaling (MDS) techniques are applied when for a set of observed similarities (or distances) between every pair of N items, we want to find a representation of the items in fewer dimensions such that the inter-item proximities nearly match the original similarities (or distances). It may not be possible to match exactly the ordering of the original similarities (distances). Consequently, scaling techniques attempt to find configurations in $q \leq N - 1$ dimensions such that the match is as close as possible. The numerical measure of closeness is called the *stress*. [30] summarizes the MDS algorithm as follows:

- For N items, obtain

  $$M = \frac{N(N-1)}{2} \tag{15}$$

  similarities (distances) between distinct pairs of items.

- Order the similarities as

  $$s_{i_1k_1} < s_{i_2k_2} < \ldots < s_{i_Mk_M} \tag{16}$$

  where $s_{i_1k_1}$ is the smallest of the M similarities.

- Using a trial configuration in q dimensions, determine the inter-item distances

  $$d_{i_1k_1}^{(q)} > d_{i_2k_2}^{(q)} > \ldots > d_{i_Mk_M}^{(q)} \tag{17}$$

- Minimize the *stress*

$$Stress = \left[\frac{\sum_{i<k} \sum (d_{ik}^{(q)} - \hat{d}_{ik}^{(q)})^2}{\sum_{i<k} \sum (d_{ik}^{(q)})^2} \right]^{\frac{1}{2}} \tag{18}$$

- Using the $\hat{d}_{ik}^{(q)}$'s, move the points around to obtain an improved configuration. A new configuration will have new $d_{ik}^{(q)}$'s new $\hat{d}_{ik}^{(q)}$'s and smaller stress. The process is repeated until the best (minimum stress) representation is obtained.

Thus, MDS allows to visualize how near points are to each other for many kinds of distance or dissimilarity measures and can produce a representation of data in a small number of dimensions. MDS does not require raw data, but only a matrix of pairwise distances or dissimilarities. A matrix is a similarity matrix if larger numbers indicate more similarity between items, rather than fewer. A matrix is a dissimilarity matrix if larger numbers indicate less similarity.

3.8. Percentiles Range

Percentiles are statistics that apply to numeric data to describe the data sample spread. Percentiles are based on division into 100 units. The *Pth* divide the data such that there is a P percentage of the numbers below the *Pth* percentile and a percentage $100 - P$ above the *Pth* percentile. The way to find the *Pth* percentile is shown as follows:

- Sort all observations in ascending order

- Compute the position

$$L = \frac{P}{100} * N \tag{19}$$

  where N is the total number of observations.

- if L is a whole number, the *Pth* percentile is the value midway between the L value and $L+1$.

- if L is not a whole number, change it by rounding up to the nearest integer. The value at that position is the *Pth* percentile.

The percentiles range compute the difference between two percentiles of an N sample and as the percentiles provide information about the shape of the data as well as its location and spread, this difference is a robust estimate of the spread of the data such that it is more representative than the standard deviation as an estimate of the spread of the body of the data when data does not follows a normal distribution.

3.9. Autoassociative Neural Networks

[31] proposed an Autoassociatiative Neural Network (AANN) used as a nonlinear PCA (NLPCA) to identify and remove correlations among problem variables as an aid to dimensionality reduction, visualization, and exploratory data analysis. NLPCA operates by training a feedforward neural network to perform the identity mapping, where the network inputs are reproduced at the output layer. The network contains an internal bottleneck layer (containing fewer nodes than input or output layers), which forces the network to develop a compact representation of the input data, and two additional hidden layers. Figure 5 shows the architecture of this AANN with five layers.

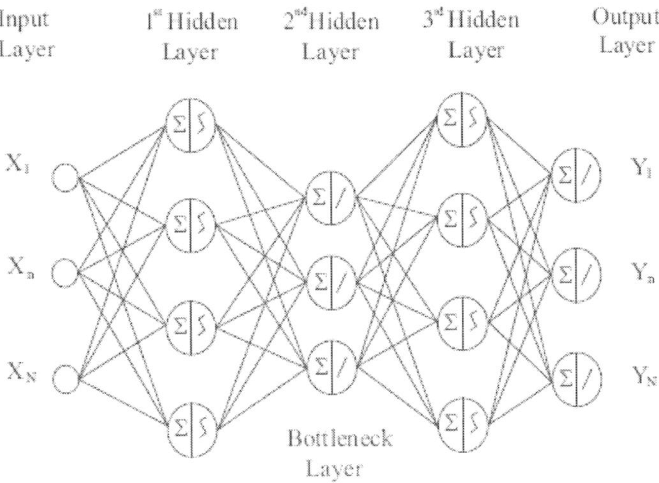

Figure 5. Architecture of an AANN.

This AANN has one input layer and one output layer each with N neurons and three hiden layers with H_1, H_2 and H_3 neurons respectively. When an observation x is presented at the input of the net, the output neurons equations of the lth layer in function of the neurons of the $(l-1)^{th}$ (see figure 6) are given by

$$y_j = f(\sum_{i=1}^{n(l-1)} w_{ji}^{(l)} y_i^{(l-1)}) \quad l = 1, ..., 4, \tag{20}$$

where $y_n^{(0)} = x_n$, $n = 1, ..., N$, represents the components of the observations vector at the input of the net. $y_n^{(4)}$, $n = 1, ..., N$, represents the components of the estimation $\hat{x}$ given at the output of the net. $n_{(l-1)}$ gives the number of neurons at the $(l-1)$ layer. $f(.)$ is the neurons activation function.

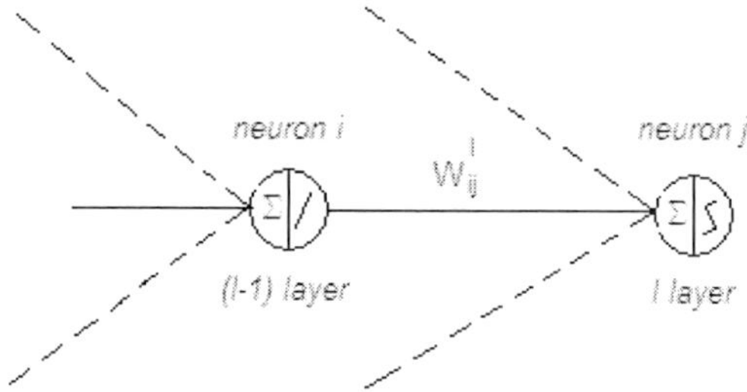

Figure 6. Conection of the ith neuron of the $(l-1)$ layer with the jth neuron of the l layer.

As the neurons activation functions are sigmoidals, the observations must be normalized as they lay inside of a unitary hypercube.

When a task is needed to be learned by the net, it is necessary to adjust the weights of the connections between neurons in order to minimize the difference between the output expected and the output given by the AANN. This minimization is carried out when calculating the derivative of the error criteria. The method commonly used is the backpropagation of the gradient error.

3.9.1. Gradient of the Partial Error Criteria

Let $J(W)$ be the partial error criteria defined as the square of the difference between the desired output x which is the input of the net too and the output estimated by the AANN y:

$$J(W) = \frac{1}{2}||x(t) - y(t)||^2 = \frac{1}{2}\sum_{n-1}^{N}(x_n - y_n)^2 \tag{21}$$

The weight $w_ji^{(l)}$ adaptation rule consists in the calculation of the sensibility of the error criteria $J(W)$ of the net

$$w^l_{ji}(t+1) = w^{(l)}_{ji}(t) - \eta\frac{\partial J(W)}{\partial w^{(l)}_{ji}} \tag{22}$$

The term $\frac{\partial J(W)}{\partial w^{(l)}_{ji}}$ is given by applying the gradient backpropagation rule to equations 20 and 21 [34]. With 20 and 22 it can be obtained the so called generalized delta rule.

$$w^{(l)}_{ji}(t+1) = w^{(l)}_{ji}(t) + \eta(t)\delta^{(t)}_j(t)y^{(l-1)}_j(t) \tag{23}$$

thus, for the weights at the output layer $l = 4$

$$\delta^{(4)}_n(t) = (x_n(t) - y_n(t))f'(y^{(4)}_n) \qquad n = 1,...,N \tag{24}$$

and for the weights at the output of one of the hidden layers $l, l < 4$

$$\delta^{(t)}_n = f'(y^{(4)}_n)\sum_{k=1}^{n_{(l-1)}}\delta^{(l+1)}_k(t)w^{(l+1)}_{kj}(t) \tag{25}$$

3.9.2. Learning Algorithm

1. Initialize the weights of the net randomly with values near 0. Fix the learning coefficient ψ and the the maximum iterations number.
2. Choose in a random way an observation x of the learning database and present it to the net
3. Backprogation: Compute the different outputs of the net given by equation 20
4. Update the net weights
5. Increment t $(t = t+1)$ and go to 2 until the error is feasible or the maximum number of iterations is reached.

3.10. Adaptive Neuro-Fuzzy Inference Systems

Adaptive Neuro-Fuzzy Inference Systems (ANFIS) are a class of adaptive networks that are functionally equivalent to fuzzy inference systems. For simplicity, assume that the fuzzy inference system under consideration has two inputs x and y and one output z. For a first order Sugeno fuzzy model a common rule set with two fuzzy if-then rules is of the form

- Rule 1: If x is A_1 and y is B_1, then $f_1 = p_1x + q_1y + r_1$
- Rule 2: If x is A_2 and y is B_2, then $f_2 = p_2x + q_2y + r_2$

and is shown in figure 7.

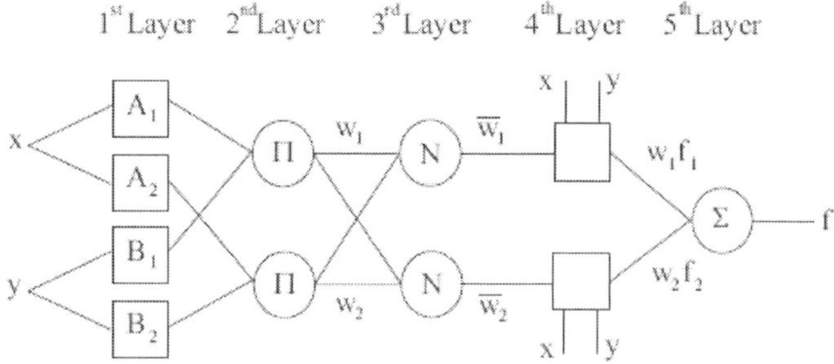

Figure 7. ANFIS architecture.

Every node i in layer 1 has a node function

$$O_{1,i} = \mu_{A_i}(x) \quad for\ i = 1,2, \tag{26}$$

where $O_{1,i}$ is the membership grade of a fuzzy set A ($A = A_1, A_2, B_1$ or B_2) and it specifies the degree in which the given input x satisfies the quantifier A. These are the premise parameters. In layer 2 every node is a fixed node labeled Π, whose output is the product of all the incoming signals

$$O_{2,i} = w_i = \mu A_i(x)\mu B_i(y) \quad i = 1,2 \tag{27}$$

each node output represents the firing strength of a rule. Every node in layer 3 is a fixed node labeled N. The ith node calculates the ratio of the ith rule's firing strength to the sum of all rules' firing strengths

$$O_{3,i} = \bar{w_i} = \frac{w_i}{w_1 + w_2} \quad i = 1,2 \tag{28}$$

In layer 4 every node i is an adaptive node with a node function

$$O_{4,i} = \bar{w}_i f_i = \bar{w}_i (p_i x + q_i y + r_i) \tag{29}$$

these are the consequent parameters. The single node in layer 5 is a fixed node labeled $\sum$, which computes the overall output as the summation of all incoming signals

$$overall\ output = O_{5,i} = \sum_i \bar{w}_i f_i = \frac{\sum_i w_i f_i}{\sum_i w_i} \tag{30}$$

4. Diagnosis Frameworks Proposed

In this section we expose several frameworks to do a complete diagnosis system and we show how each proposal works with a specific case study. All the following frameworks use a combination of the mathematical tools depicted in section 3. The order of appearance of the frameworks is the order in which they have been proposed during our research work.

4.1. Fault Detection Combining PCA, Control Charts and a Comparison With Statistic Limits

4.1.1. Framework Description

Figure 8 shows the proposed detection framework, that is a History Based fault detection method, which only requires a big quantity of historical data process. This data set takes into account only the normal system data operation since it will be transformed by both, PCA model and the normal operation data limits to be used as prior knowledge of the system to carry out the detection process.

Thus, the first step is to obtain a normal operation data set from system and a standardization is carried out. From this data set a PCA model is obtained to see the relationships and to find out the correlations between variables that at a first sight, it would be very difficult to notice. Given thus a decrement in original data matrix dimension, allowing to work with less but enough data to describe the maximum variability of original system. PCA model gives the loads and scores of the principal components corresponding to normal operation. With this loads and scores of normal operation, the value of the explained variance for each component as well as the boundaries for Hotelling's T^2 and

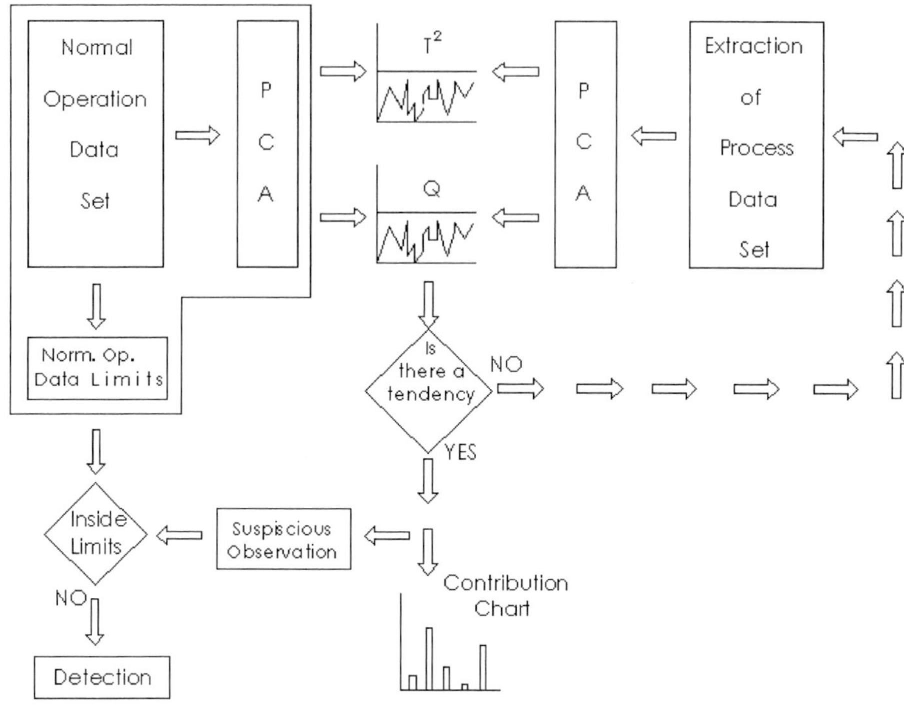

Figure 8. General fault detection framework.

Q statistics are obtained (equations 4 and 6 respectively), serving as the corresponding limits in control charts. In addition, the normal operation data limits are obtained following the steps mentioned in section 3.3.. An modification in method depicted in [11] is been done. Instead of taking several signal transformation functions (STF) and so many single output functions (SOF), here it is taken just one STF and one SOF. Raw data is taken as the STF and the mean as the SOF, making it possible to obtain the minimum and maximum statistic limits that variables being monitored should have in normal operation.

For detection process, a process data set is extracted and analyzed as follows. A PCA model is build from the extracted process data set, and the scores of the principal components given by this model are plotted in control charts comparing them with the T^2 and Q statistics corresponding to normal operation model. If the chart does not exhibits a tendency, another process data set is extracted to be analyzed. But if a normal operation T^2 or Q limit is violated or a tendency is present, two actions are taken. 1) A contribution chart is done to find out which variable or variables has the major contribution to system's variability, and 2) the suspicious observation is compared to the normal operation data limits previously obtained to verify if it is between them or not.

4.1.2. Case Study

This section shows the performance of the framework proposed in a simulation example. The simulation consists in the operation of a system (machine) formed by five subsystems. Each subsystem is simulated with different RL series circuits (see figure 9). A change within $\pm 10\%$ of the original values in each subsystem's components is considered as normal operation. An electrical current sensor is available such that each subsystem's current could be measured.

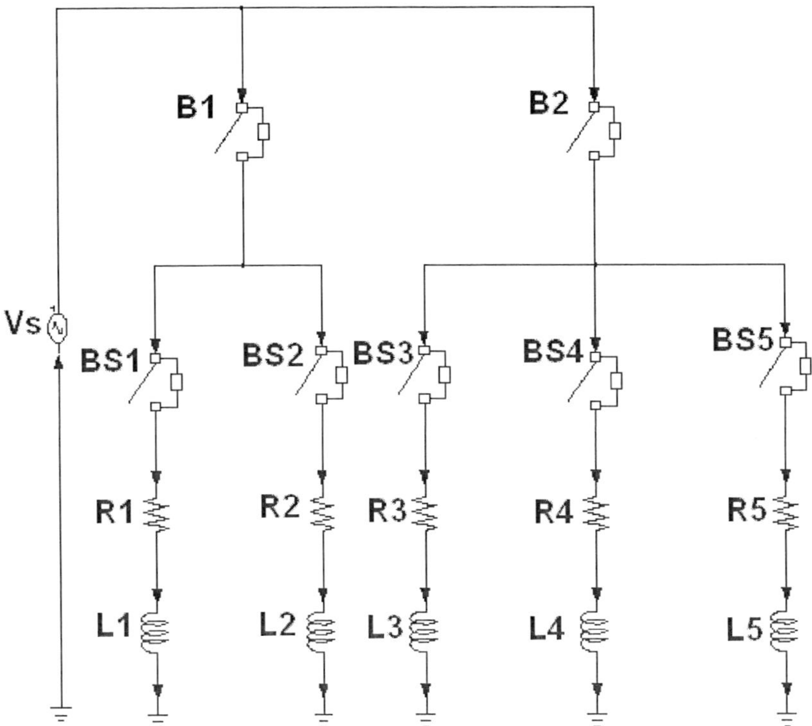

Figure 9. Simulation of a system formed by five subsystems.

Thus, methodology proposed is applied as follows:

1. From normal operation history process data (electrical current in each subsystem), build PCA model and obtain T^2 and Q statistics as well as the minimum and maximum limits for each subsystem's current.
2. Take a probe data set.
3. Apply PCA and obtain a reduction on original dimensions.
4. Build and observe control charts for T^2 and Q statistics. If control chart detects a tendency in a specific time instant go to 5), else go back to 2).
5. Build a contribution chart and obtain the electrical current value for suspicious time instant (sample).
6. Compare the suspicious electrical current value with its normal operation limits obtained above in step 1 and detect which subsystem is in faulty mode

4.1.3. Single Fault

Several runs have been done to detect simple faults and obtain the effectiveness percentage in this task. A ramp, a step and a combination of both were simulated in different subsystems to see the performance of the methodology proposed. For instance, a simulation of a single fault present in subsystem 2 in which current decrements in steps of 0.9% is included in sample 80. Figure 25 shows how control charts depict a tendency to pass its control limits.

Figure 11 shows that variable 2 (subsystem 2) is the one that has the major variability of the system, indicating thus that subsystem 2 is probably in faulty mode. Then, when checking this suspicious observations with its corresponding normal operation limits, it is found that subsystem's 2 electrical current value has decreased above the lower current normal operation limit, having in this way the detection of the simulated fault.

4.1.4. Multiple Fault

In the case of multiple faults, several runs have been carried out simulating a fault in two subsystems simultaneously. A ramp in both subsystems, a ramp in one and a step in the other, a step in both, and a combination of a negative and

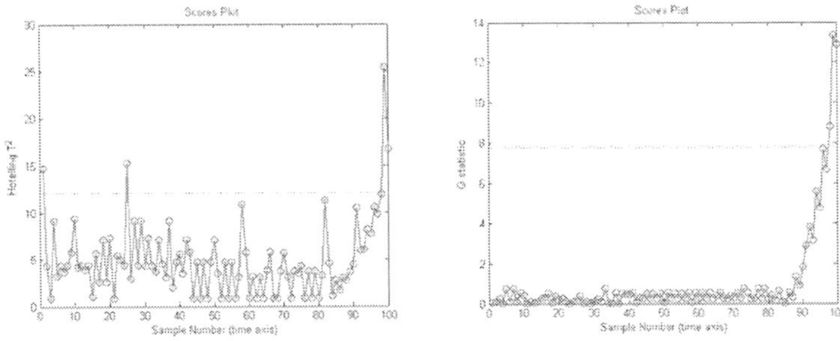

Figure 10. Control charts for a decrement of 0.9% in subsystem's 2 current.

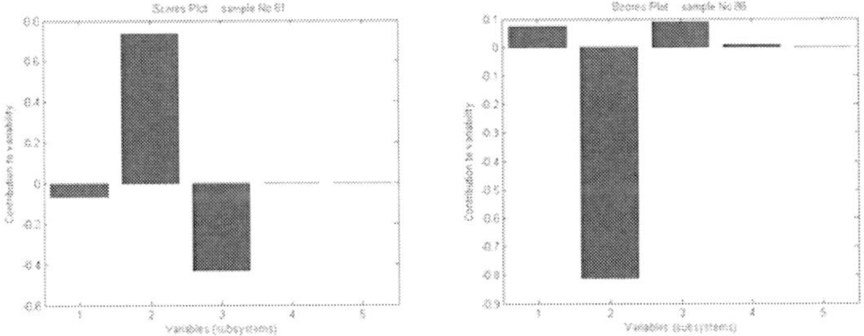

Figure 11. Contribution charts for samples 81 and 86, indicating a probably problem in subsystem 2.

positive ramp and step in one, the other or both. Figure 12 shows how control charts depict again a tendency to pass its control limits.

But a contribution chart not always shows the real variables that possibly have problems. That is why the use of minimum and maximum limits for normal operation data is important. As probably faulty subsystems are not detected by contribution charts they are truly detected by the comparison of the subsystem's electrical current suspicious value with its corresponding limits.

4.1.5. Measurement Noise

Table 1 shows the results for simulations taking into account different measurement noise magnitudes present in one, two or three subsystems. Figure 13 depicts the behavior of control charts when it exists measurement noise of

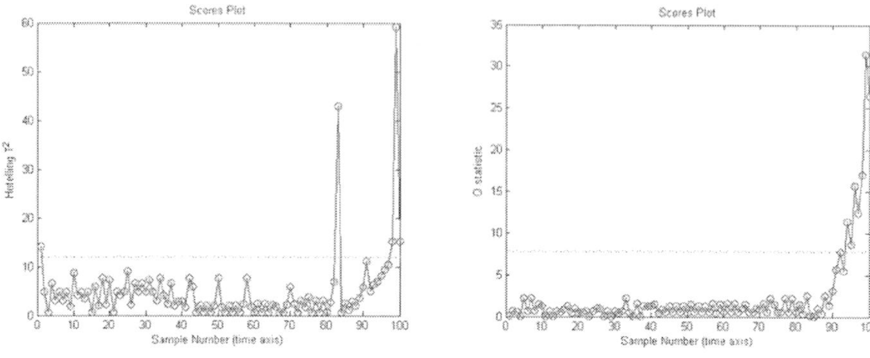

Figure 12. Control charts for an increment of 10% in subsystem 3 and a decrement of 0.9% in subsystem's 5 currents.

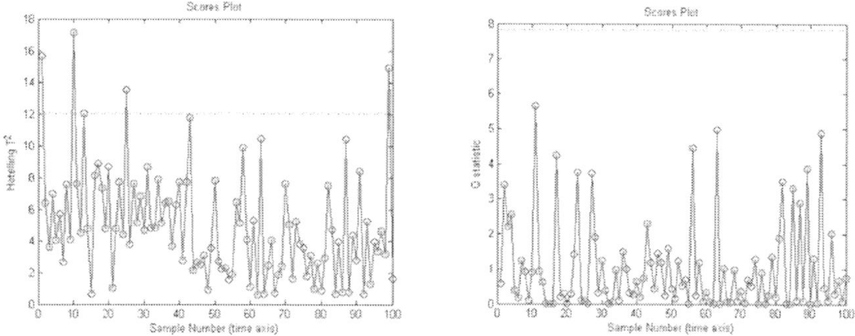

Figure 13. Presence of measurement noise of magnitude 0.1 in subsystem 1.

magnitude 0.1 present in subsystem 1.

It is important to note that none of the control charts shown in figure 13 has a specific tendency. Samples above T^2 control chart limit are outliers. Note that all samples in Q control chart are below its limit, which has sense because in this case noise does not brake the original correlation between variables. As T^2 is a measure of the multivariate distance between samples with respect to the center of data, it could detect a fault that keeps the correlation structure which could not be detected by Q. Q detects faults that violate mass or energy balance pointing out a correlation breakdown.

Table 1. Percentage detected of measurement noise present in one, two and three subsystems

Noise Magnitude	One Subs.	Two Subs.	Three Subs.
0.0001	100%	100%	100%
0.001	100%	100%	85%
0.01	95%	85%	80%
0.1	90%	85%	75%

4.1.6. Summary of the Framework

This proposal is easy to implement and to adapt because when original process changes, it is only required to modify the original data base instead of develop a new mathematical model from it. Another advantage is that the use of PCA model allows to work with less quantity of data, but keeping the original correlation between variables. It is important to note that the use of T^2 and Q control charts allows to distinguish between the presence of a fault and the presence of measurement noise. Finally the use of minimum and maximum limits for comparisons between a suspicious sample and its normal values gives the detection of a single and multiple faults presents in the system.

4.2. Multiple Fault Diagnosis in Electrical Power Systems with Probabilistic Neural Networks

4.2.1. Framework Description

The second proposed detection framework is shown in figure 14. The framework only requires a big quantity of historical data of the power system's nodes containing normal operation data and faulty data samples from the different types of faults that could be present in the system. These data sets are used as prior knowledge of the power system to perform the detection process.

The first step is to obtain several data sets from the power system's nodes. These data sets are matrices formed by windows of m samples and n power system's nodes where the voltage of each line of a certain node is monitored, that means three readings per node. Such matrices are constructed with normal node operation and different node faults present in system. For each node data sets its correlation matrix is obtained to see how their three lines are related. Once

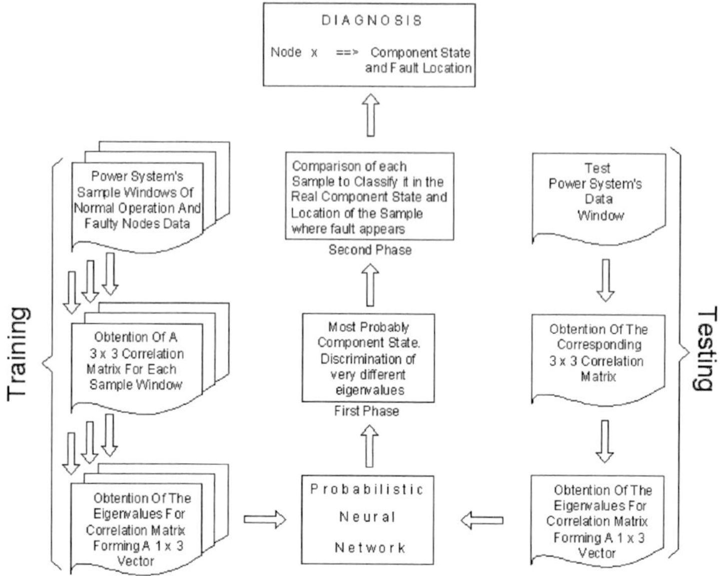

Figure 14. General fault detection framework.

having the correlation matrix, their corresponding eigenvalues are computed as shown in subsection 3.6., such that in this way a signature to each of the different possible component states or fault types of the power system's nodes (K in figure 4) are given. The eigenvalues of the correlation matrix for each fault signature, then serve as the training vectors of the PNN corresponding to Q as described in subsection 3.3.. Each node will have then three eigenvalues (R components in figure 4) as they are coming from its correlation matrix that is a 3×3 matrix. Then the detection process is carried out in two phases. First phase, that serves as a first filter or information discriminator. When there is a system to monitor, a window of m samples and n power system's nodes is taken, then each node is analyzed separately. m samples of the three lines corresponding to a node being monitored to find out which is its most probably state are taken. From the data set corresponding to a node being monitored its correlation matrix and its corresponding eigenvalues are obtained and then these last are used as an input vector to the PNN previously trained. It is mentioned "the most probably state" because unfortunately not all the eigenvalues of all of the node states are so different such that PNN could not classify them easily. But it has been found that for certain signature faults eigenvalues are very similar, thus there is here a discrimination/classification phase, because it is necessary to look for the real state but only comparing between just a couple of similar signatures instead of the whole bunch of node states. The output from the PNN automatically discriminates node states that are very different and gives the most probably real node state. Once the one possible node state is obtained, a second phase of the framework begins to work. In the second phase each sample of each node is taken and its magnitude is obtained, then a comparison against a constant magnitude of the probably signature faults are carried out. This comparison serves as a classifier that gives the real node component state and can be used too to locate the period of time or sample number where the fault occurs.

4.2.2. Case Study

This subsection shows the performance of the second framework proposed for multiple-fault diagnosis over the IEEE network shown in figure 15. This is an electrical power system having dynamic load changes. 24 fault simulations were carried out to determine the performance of the approach including in them symmetrical and unsymmetrical faults at random nodes (3,9,10 and 13) taking into account different multiple faults scenarios combining faults such as: one line to ground (A GND), two lines to ground (A-B GND), three lines to ground (A-B-C GND), or faults between two lines (A-B or B-C) and the no fault mode (NO FAULT).

The methodology proposed is tailored as follows:

1. Obtain windows of 100 samples from normal and faulty operation history process data (electrical voltage in each node's line).
2. Obtain correlation matrix for each node, which gives a 3×3 matrix.
3. Obtain the eigenvalues from the correlation matrix (this gives 3 eigenvalues), with this 3 eigenvalues form an input vector to train the PNN.
4. Take a test data set of 100 samples from the electrical power system being monitored.
5. Obtain correlation matrix for each node, which gives a 3×3 matrix.
6. Obtain the eigenvalues from the correlation matrix (this gives 3 eigenvalues), with this 3 eigenvalues form an input vector to the PNN.

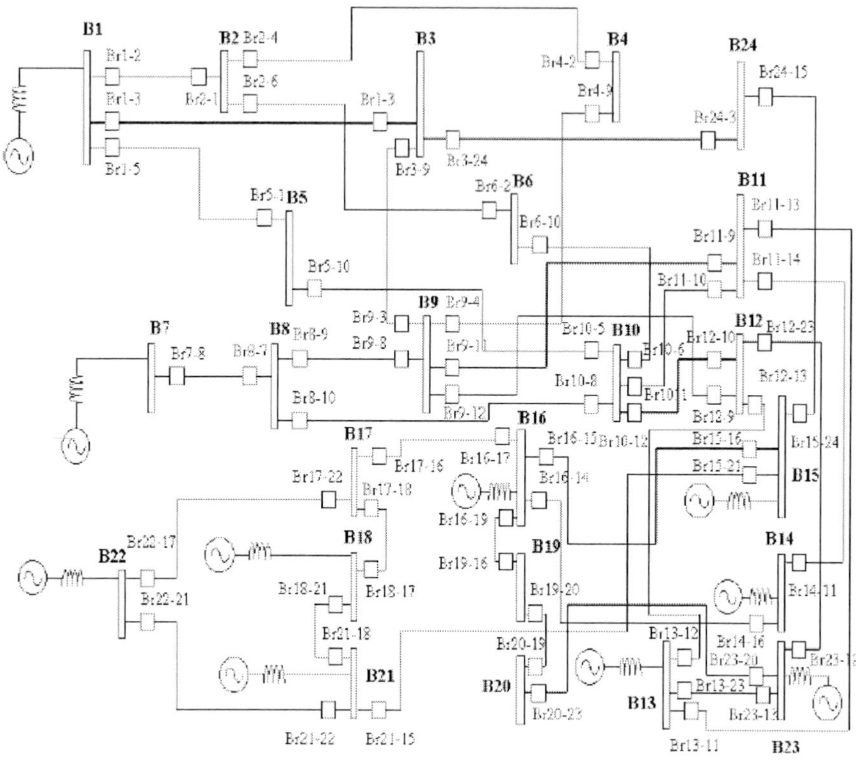

Figure 15. IEEE reliability test system single line diagram.

7. First Phase: Take the output of the PNN as one of the two probably states of the node monitored.

8. Second Phase: Take each sample of each node monitored and obtain its magnitude, then compare it against the constant magnitude of the two probably signature faults and classify it using this simple criteria. Locate the period of time or sample number where the fault occurs.

9. Give the diagnosis of each node being monitored. If a fault is present in a specific node give the type and location of it, else print NO FAULT.

In the following tables the performance of the approach is shown taking into account three possible cases. Case 1, system is working properly during the first 25 samples from a total of 100, that means 25 samples are ok and 75 samples corresponds to fault present in system. Case 2 takes 50 samples of normal operation data and 50 samples with fault present. And case 3 takes 75 samples of normal operation and 25 with fault present. Table 2 and 3 shows a detailed example of how percentages were obtained. Tables 4 and 5 gives a summary of the obtained percentages for each of the three cases considered.

Table 2. Detail performance of detection per node's component state with 25 samples ok and 75 samples with fault present (case 1)

Component State	Correct	Wrong	Accuracy
A-B-C GND	14	0	100%
A-B GND	10	0	100%
A GND	14	0	100%
A-B	18	0	100%
B-C	16	0	100%
NO FAULT	13	11	54.16%

Table 3. Detail performance of detection per node number with 25 samples ok and 75 samples with fault present (case 1)

Node Number	Correct	Wrong	Accuracy
3	20	4	83.33%
9	19	5	79.16%
10	22	2	91.66%
13	24	0	100%

Figure 16 shows an example of the results obtained for a simulation of case 2, that has the following faults types:

- 3 A GND, that is a fault present in node 3 of type line A to ground.
- 9 A-B GND, that is a fault present in node 9 of type lines A and B to ground.
- {10,13} NO FAULT, that is nodes 10 and 13 working properly.

```
>> Eigenvalues of correlation matrix from node 3

eigenvalues1 =

    1.4936    1.3614    0.1450

Eigenvalues of correlation matrix from node 9

eigenvalues2 =

    1.4970    1.3143    0.1887

Eigenvalues of correlation matrix from node 10

eigenvalues3 =

    1.5276    1.4615    0.0109

Eigenvalues of correlation matrix from node 13

eigenvalues4 =

    1.5143    1.4857    0.0000

Fault present in node 3 type  ==>   A GND beginning from sample number 52
Fault present in node 9 type  ==>   A-B GND beginning from sample number 52
Node 10  ==>  NO FAULT
Node 13  ==>  NO FAULT
>>
```

Figure 16. Example of the results given by a simulation software.

It has been found out that for the test power system being analyzed, eigenvalues corresponding to A-B-C GND were very similar to those corresponding to NO FAULT, in the same way eigenvalues corresponding to A GND fault and A-B GND fault as well as A-B fault and B-C fault were very similar respectively, addressing us to implement the second phase of the framework to tackle this problem and to achieve a good performance in the diagnostic.

From figure 16, it is important to note that does not matter that the eigenvalues for each of the nodes are very similar because, as said before, in the first phase of the framework a first classification/discrimination is done and in the second phase a second classification and location of the fault is carried out.

From tables 4 and 5 it can be said that the different results in performance are due to eigenvalues of correlation matrices with very similar values, when

Table 4. Accuracy of detection per node's component state for the different cases

Component State	Case 1	Case 2	Case 3
A-B-C GND	100%	100%	100%
A-B GND	100%	100%	100%
A GND	100%	85.71%	92.85%
A-B	100%	83.33%	50%
B-C	100%	68.75%	68.75%
NO FAULT	54.16%	58.33%	79.16%

Table 5. Accuracy of detection per node number for the different cases

Node Number	Case 1	Case 2	Case 3
3	83.33%	83.33%	83.33%
9	79.16%	75%	70.83%
10	91.66%	87.5%	62.5%
13	100%	95.83%	100%

there are a major quantity of normal operation data in sample window. When more normal operation data appear in the sample window, more difficult is to classify eigenvalues by PNN because they are very similar.

An important thing to remark is that there has been carried out several tests when all data were from normal operation and it has been found out that the framework have detected 100% of them as NO FAULT node's component state. Percentages shown in tables are low because criteria used in the second phase of the framework is related with the maximum magnitude value and a threshold that needs to be set as the upper limit to make the difference between two very similar signatures for the same node.

4.2.3. Comparison Against a Diagnostic System Based on Probabilistic Logic

To observe the general performance of the proposal, a comparison against a diagnostic system based on probabilistic logic has been carried out. Table 6 and 7 shows the performance of the diagnostic system based on probabilistic logic.

Table 6. Performance of detection per node's component of the diagnostic system based on probabilistic logic

Component State	Correct	Wrong	Accuracy
A-B-C GND	14	0	100%
A-B GND	10	0	100%
A GND	12	2	85.7%
A-B	15	3	83.3%
B-C	16	0	100%
NO FAULT	17	7	70.8%

Table 7. Performance of detection per node number of the diagnostic system based on probabilistic logic

Node Number	Correct	Wrong	Accuracy
3	19	5	79.1%
9	21	3	87.5%
10	21	3	87.5%
13	23	1	95.8%

Comparing the results of both frameworks it can be noted that in general they have a very similar performance, but when comparing case 1 of this second proposal against the diagnostic system based on probabilistic logic it could be said that our methodology has a better performance. Another important point is that our proposal is relatively easier to implement and to update than that based on probabilistic logic when power system grows up.

4.2.4. Summary of the Framework

We have presented a fault detection framework for electrical power systems with dynamic load changes using a PNN based on history process data. An advantage over model based methods is that this framework needs historical data of normal system operation as well as faulty data sets to train the PNN. It has been decided to use the PNN because is ideal when working on classification problems. Its most important advantage is that it needs only a little time for its training. It has been proposed a two phases relatively easy to implement fault diagnosis method. In the first phase the eigenvalues of the correlation matrix are taken and used them as inputs for a PNN to classify the node's component state. It has been shown how this classification could be improved and carry out when eigenvalues are very similar with the implementation of a second phase where a simple comparison of each sample magnitude to the constant value of a certain signature fault has been applied and at the same time it gives the location of a fault if it is present. It can be concluded too, that as there is as many faulty data in the sample window the proposal has a better performance because eigenvalues are easily classified by the PNN as they have very different values. Another advantage of this proposal is that as it diagnostics each node, it could be detected simultaneous and nonsimultaneous simple, multiples, a combination of different faults as well as their corresponding location on each node separately.

4.3. Fault Detection and Diagnosis of a Vehicle Using Soft Computing over the Residuals of a Model Based Method

In this subsection we propose a fault detection framework combining a vehicle model based method and soft computing applied over the residuals to carry out the diagnosis. We have chosen a vehicle as the system to be monitored because a vehicle is a highly complex system and then is one of the most challenging systems to monitor. The combination of methods is done in order to reduce the number of false alarms in the diagnosis process as well as to deal with the presence of high measurement noise. The framework proposed is divided in two phases. In the first phase a vehicle model described in [51] is used to obtain a set of residuals. The model considered here consists of 7 degrees of freedom (DOF), which include longitudinal and lateral motions, yaw motion and the rotational dynamics of the four wheels. In the second phase a vector is formed with the residuals and then a comparison against their normal operation limits using fuzzy logic is done to look for those residuals standing out

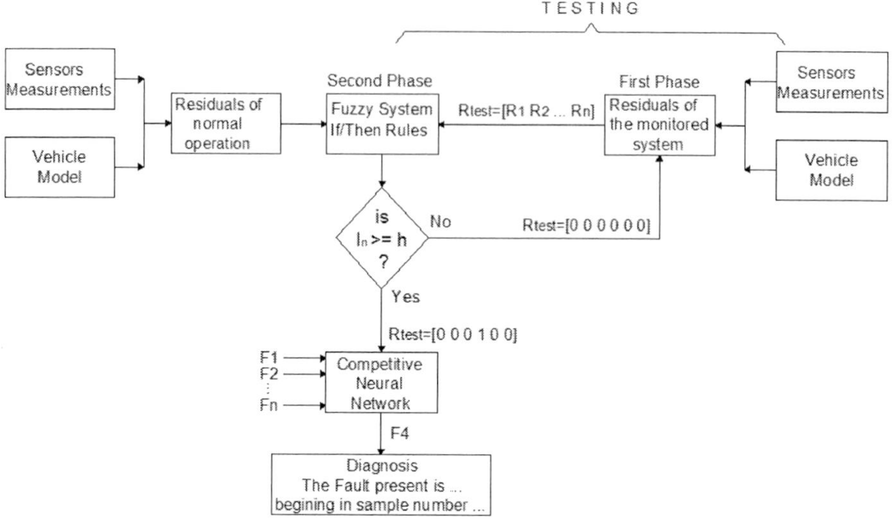

Figure 17. General fault detection and diagnosis framework.

of limits. Then a competitive neural network classifies the output of the fuzzy system indicating in this way which fault is present on the system. We focus this presentation on the second phase of the framework as it is the one that uses the soft computing methods in order to carry out the detection and diagnosis of a fault.

4.3.1. Framework Description

The general fault detection and diagnosis framework is shown in Fig. 17 and it works as follows. First a data set of the residuals coming from the normal operation of the system is required. Then the minimum, maximum and statistical mean values for each of these residuals are obtained. The rules of the fuzzy system can be stated with these statistical parameters, and they will be used to check if a residual is out of its normal operation limits. In this work, we use the gaussian membership function shown in (31) for the fuzzy system [33] [34].

$$gaussian(x; c, \sigma) = e^{\frac{1}{2}\left(\frac{x-c}{\sigma}\right)^2} \tag{31}$$

And the set of fuzzy rules used are the ones shown below

- If value is inside limits then R_n is zero
- If value is outside limits then R_n is one

Where R_n is the residual from the nth output of the model and the nth sensor, and the current values for inside and outside limits were those obtained with the minimum, maximum and statistical mean computed from the set of residuals for normal operation of the system.

The first phase gives the residuals of each of the vehicle model and measurement sensors in a vector R_{test} of dimension $1 \times n$ formed with these n residuals as follows:

$$R_{test} = [R_1, ..., R_n] \tag{32}$$

This R_{test} vector is analyzed by the fuzzy system in order to obtain which residuals are out of its limits. If this is the case, the fuzzy system assigns this residual a value of one, otherwise, when residuals are inside their limits, they are indicated with a zero. In order to detect false alarms, an index must be defined to monitor the number of consecutive points, given by the fuzzy system, that lay out of their limits given. In this way we can distinguish between noise presence or the existence of a real fault. Thus, this index I_n will count the consecutive number of times or points the residual R_n has out of its limits, such than when I_n reaches h that is a constant number set by the user, a modified R_{test} vector will be formed as that shown in (33), that shows for instance the R_{test} modified when R_4 of an R_{test} vector for a system formed by six residuals has reached h consecutive points out of its limits.

$$R_{test} = [0\ 0\ 0\ 1\ 0\ 0] \tag{33}$$

Then this vector serves as the input of the competitive neural network previously trained in order to be classified. Thus the example shown in (33) could indicate a fault present in the right rear wheel sensor for instance. As this process is carried out over each of the R_n residuals forming the R_{test} vector obtained from the model based method, it is possible to know in which time or sample number a fault starts.

The algorithm of the second phase could be summarized as follows:

1. Obtain a set of residuals from a normal operation data set.
2. Obtain the minimum, maximum and mean values for each residual of the set obtained of normal operation data from step 1

3. Build the fuzzy system for each residual of the system being monitored according with the values obtained in step 2 with the if / then rules of the form

   - If value is inside limits then R_n is zero
   - If value is outside limits then R_n is one

4. Define an index I_n for each residual in order to distinguish between noise presence and a fault present. $If\ I_n \geq h$ then a fault is present on R_n $ELSE$ measurement noise is present on R_n. h is the number of consecutive points out of the limit allowed and it is set by the user.

5. Learn a competitive neural network with the different fault signatures in order to classify the output of the fuzzy system.

Thus, when there is a system to monitor it will work as described below:

6. Build a residuals vector to test of the form

   $R_{test} = [R_1, \ldots, R_n]$

7. Take the R_{test} vector as the input of the fuzzy system. The output of it is a modified R_{test} binary vector composed with 0s and 1s. Thus, it will have a 0 if the residual is inside its normal operation limits or a 1 if it is out of them.

8. Verify the index of each residual I_n. If the points laying out of limits are consecutives *then* increment the index I_n and *Go* to step 9 $ELSE$ reset I_n to zero.

9. $If\ I_n \geq h$ *then* stop the monitoring process and *Go* to step 10 $ELSE$ return to step 6.

10. Put the R_{test} vector obtained in step 7 as the input of the probabilistic neural network previously trained, in order to carry out the diagnosis process of the system being monitored. Give which fault is present and its location.

4.3.2. Case Study

This subsection gives some results in order to test and validate the proposed method. We have performed several experiments, where the state and forces were generated by using the vehicle simulator VEDYNA [35]. We consider six variables to monitor:

1. Left front wheel angular velocity (LFW)
2. Right front wheel angular velocity (RFW)
3. Left rear wheel angular velocity (LRW)
4. Right rear wheel angular velocity (RRW)
5. Longitudinal velocity of the center of gravity (Vx)
6. Yaw velocity (Yaw)

in two different vehicle dynamics: chicane and slalom and three different databases:

1. Normal operation residuals for chicane database
2. Fault present in residual R1 chicane database
3. Fault present in residual R4 slalom database

Fig. 18 shows the output of the fuzzy system for the six residuals when vehicle works in normal operation for chicane database.

As mentioned above the index defined in step 4 of the algorithm is used to distinguish between noise or fault present on a system. Thus when monitoring a system which does not have either a fault or noise presence this index will never reach h consecutive points for a certain R_n. On the other hand, when monitoring a system that has a point out of its normal operation limits the fuzzy system will give to it a 1 and I_n increases in one. If there are h consecutive points out of their normal operation limits for a residual R_n the existence of a fault on it must be expected on the final diagnosis. The index I_n increases when consecutive outputs of the fuzzy system are 1, and resets to zero when the consecutives outputs of the fuzzy systems are different form 1. Thus, when

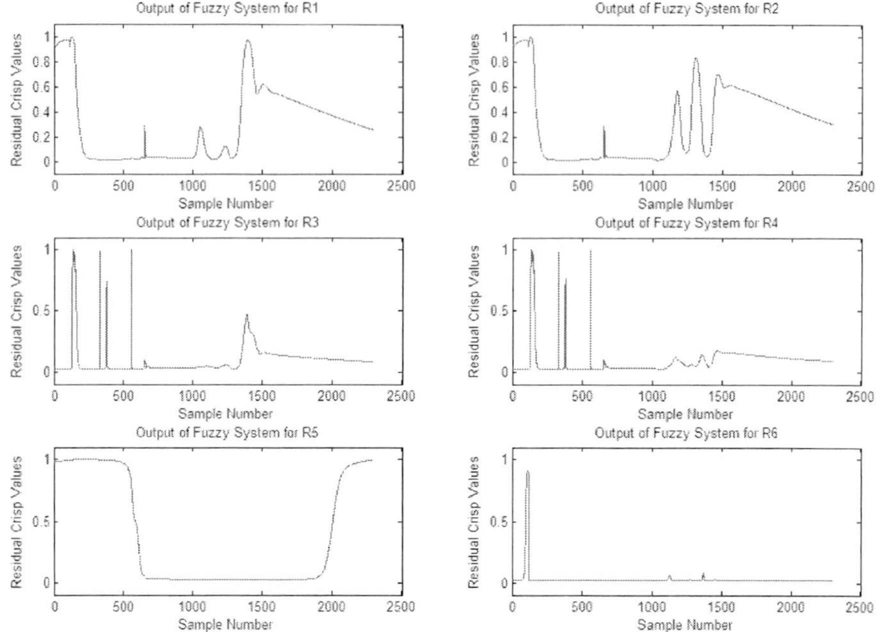

Figure 18. Defuzzyfied output of the fuzzy system for normal operation residuals for chicane database.

monitoring a system having several points out of their limits without reach h consecutive points, they are considered as measurement noise presence.

When there is a fault present on a certain residual, the output of the fuzzy system is taken as the input of the probabilistic neural network in order to determine the type of fault. As this process is carried over each residual when a fault exists, its location over time is giving immediately. After carry out several simulation tests it has been found out as shown in table 8 that the best performance of the proposal is reached when $h = 10$.
Fig. 19 and Fig. 20 show the results of the proposal when a fault exists on R_1 and R_4 on chicane and slalom databases respectively. In both cases, the index used to distinguish between noise or fault was $h = 10$.

The results given by the fuzzy system when a fault is present in R_1 chicane database are explained as follows:

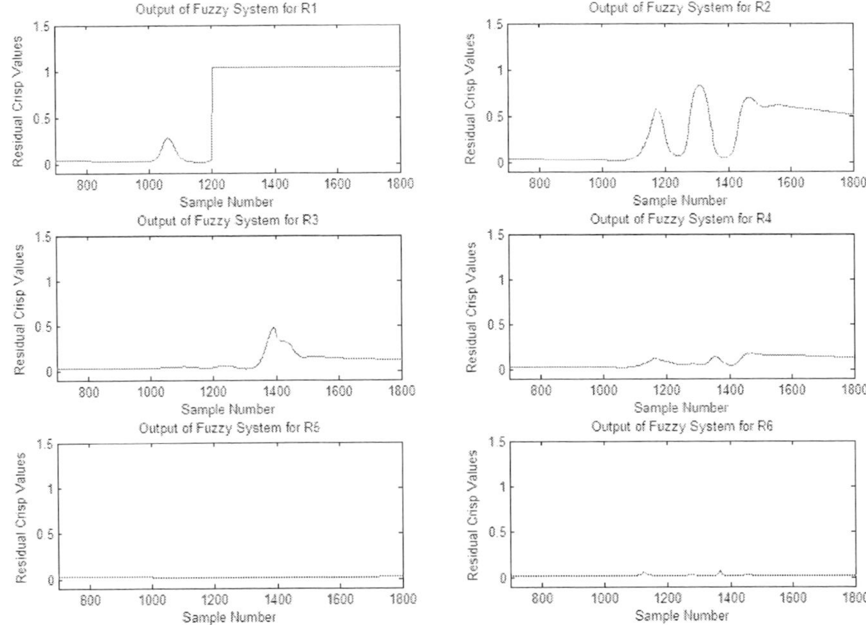

Figure 19. Fault detection for a simulation of chicane data base with fault present in R_1 without noise and using $h = 10$.

- There are 10 points out of the limits for residual R_1 and their locations are samples number 1201, 1202, 1203, 1204, 1205, 1206, 1207, 1208, 1209 and 1210.
- There are 0 points out of the upper limit for residual R_2.
- There are 0 points out of the upper limit for residual R_3.
- There are 0 points out of the upper limit for residual R_4.
- There are 0 points out of the upper limit for residual R_5.
- There are 0 points out of the upper limit for residual R_6.

As it was expected, just residual R_1 has points out of its limits because it was a simulation without measurement noise presence. In addition to this it could be noticed that the fault is found out when having $h = 10$ consecutive points out of the limits of R_1. Note that fault starts at sample 1201 that is the first of the ten consecutive points.

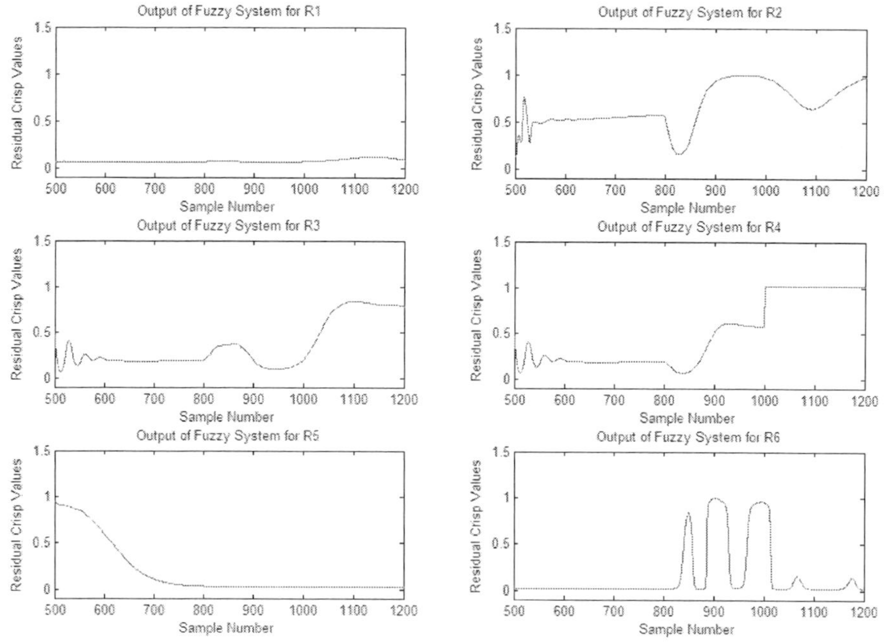

Figure 20. Fault detection for a simulation of slalom data base with fault present in R_4 without noise and using $h = 10$.

The results for fault present in R_4 slalom database simulation could be seen on figure 20 and they are as follows:

- There are 0 points out of the upper limit for residual R_1.
- There are 3 points out of the limits for residual R_2 and their locations are samples number 990, 991 and 992.
- There are 0 points out of the upper limit for residual R_3.
- There are 10 points out of the limits for residual R_4 and their locations are samples number 1000, 1001, 1002, 1003, 1004, 1005, 1006, 1007, 1008 and 1009.

- There are 0 points out of the upper limit for residual R_5.
- There are 0 points out of the upper limit for residual R_6.

As R_2 has 3 consecutive points out of the limits, they could be taken as outliers (Fig. 20). But as it was expected, just residual R_4 has $h = 10$ consecutive points out of its limits indicating thus the presence of a fault on it. Note that fault starts at sample 1000 that is the first of the ten consecutives points.

4.3.3. Measurement Noise

Table 8 shows the results for simulations when different measurement noise magnitudes are present in all the residuals simultaneously for different quantities of consecutive points (h). This table shows that when working with measurement noise presence in the system, an index of $h = 10$ is preferred to be use as it could distinguish between noise and fault present in all the simulations that were carried out. The percentages shown were obtained from 20 simulations with 2300 samples each of normal operation for each pair formed with a noise magnitude and an h value. Giving thus a total of 320 simulations.

Figure 21 depicts the behavior of the outputs from the fuzzy system for normal operation when there is measurement noise present of magnitude 0.01 and $h = 10$ in all the residuals simultaneously. The results for noise presence in chicane normal operation database simulation are as follows:

- There are 16 points out of the upper limit for residual R_1, and their locations are samples number 127, 135, 1356, 1382, 1385, 1387, 1393, 1395, 1397, 1399, 1401, 1407, 1408, 1414, 1431 and 1573.
- There are 11 points out of the upper limit for residual R_2, and their locations are samples number 129, 130, 134, 135, 1308, 1312, 1447, 1499, 1623, 1891 and 2047.
- There are 5 points out of the upper limit for residual R_3, and their locations are samples number 134, 135, 141, 148 and 558.
- There is 1 point out of the limits for residual R_4 and its location is sample number 135.
- There are 73 points out of the limits for residual R_5 and their locations are samples number 148, 149, 160, 173, 178, 187, 203, 208, 210, 227, 235,

Table 8. Performance of detection for different magnitudes of measurement noise present in all the residuals simultaneously for different h values

Noise Magnitude	$h = 10$	$h = 7$	$h = 5$	$h = 3$
0.0001	100%	100%	100%	85%
0.001	100%	100%	90%	50%
0.01	100%	95%	60%	20%
0.1	100%	70%	25%	5%

236, 237, 244, 245, 246, 250, 308, 309, 318, 2021, 2040, 2073, 2077, 2089, 2096, 2107, 2111, 2119, 2124, 2131, 2139, 2146, 2152, 2167, 2173, 2176, 2180, 2193, 2194, 2196, 2198, 2200, 2208, 2209, 2210, 2219, 2232, 2234, 2237, 2238, 2239, 2251, 2254, 2255, 2256, 2258, 2264, 2265, 2269, 2271, 2272, 2274, 2277, 2281, 2283, 2284, 2289, 2292, 2296, 2297, 2299 and 2300.

- There is 0 point out of the limits for residual R_6.

As there were not $h = 10$ consecutive points out of the limits in any of the residuals, all the nonconsecutive points out of the limits are considered as measurement noise present.

4.3.4. Summary of the Framework

This subsection has presented an online fault detection framework based on the treatment of the residuals obtained from a model based method for a vehicle diagnostic system. The main advantage of this approach is that it diminishes the number of false alarms in the diagnosis process as well as it could deal with high measurement noise presence. This framework is divided in two phases, the first of it uses a model based technique to obtain the residuals between the output of the model and the measurements. In the second phase these residuals are treated with soft computing techniques. These are a combination of fuzzy logic and a probabilistic neural network. The implementation of an index that grows up when a constant number of consecutive points standing out of limits given by the fuzzy system allows to distinguish between measurement noise or a

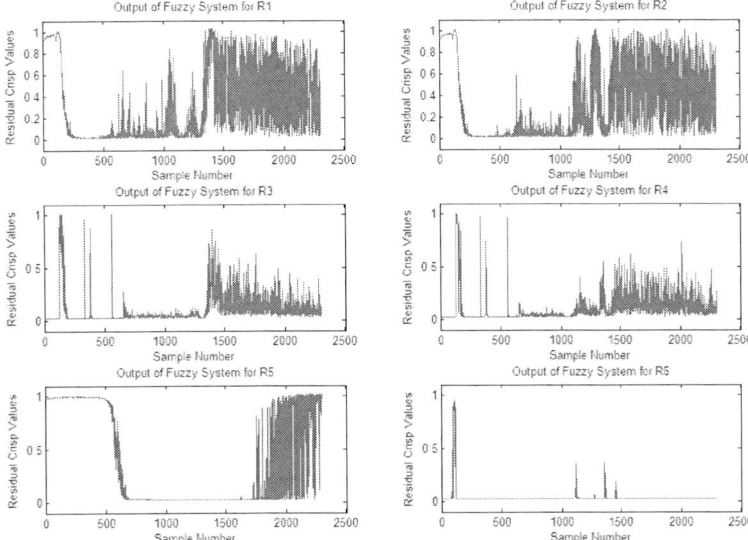

Figure 21. Presence of measurement noise of 0.01 magnitude on normal operation residuals and $h = 10$.

fault present in the system. A probabilistic neural network is used to classify the outputs of the fuzzy system in order to carry out the diagnosis. As the process diagnostics each residual individually, this approach gives as a final diagnosis which residual is in faulty mode as well as the location of the fault present.

4.4. Fault Detection and Diagnosis of a Vehicle Combining Multidimensional Scaling, Percentiles Range and PCA

4.4.1. Framework Description

The general fault detection and diagnosis framework is shown in figure 22. This proposal needs a data set of normal operation behavior of the system. Then as depicted in figure 22 a feature extraction step is performed to learn the normal operation dynamics of the system. These features are compared against others features extracted from observations of the system. The feature extraction block includes several methods to extract relevant aspects that will serve as a-priori knowledge of the normal behavior of the system. This

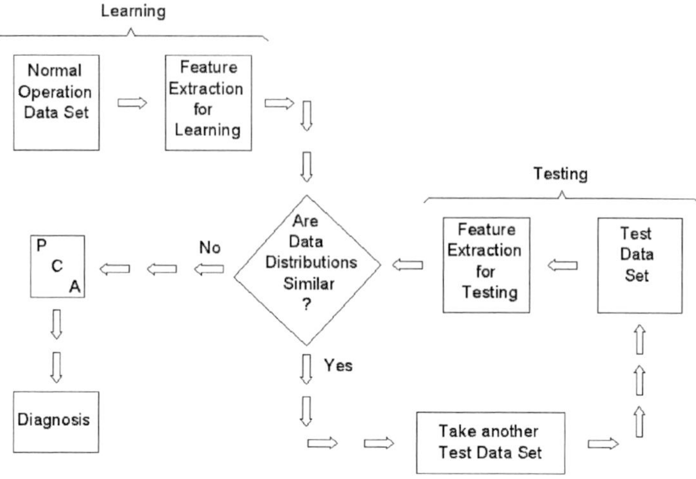

Figure 22. General fault detection and dignosis framework.

block perform its function as follows: The first step is to compute a principal components analysis (PCA) to look for the correlation between variables. This will group variables that are correlated and separate those that are not. This step is necessary to check correlation between variables that could be hard to notice, and will give the root cause of a fault. After PCA finds the correlations in the system, the data set of normal operation is split in windows of n samples. These windows are analyzed separately to have a more accurate mapping that could describe better the normal operation behavior. As it is desirable to work with less data, without missing the correlation and variability information, a multidimensional scaling (MDS) procedure is carried out. The MDS needs as an input a distance matrix D or a matrix of correlations among variables. Such matrices are categorized as either similarities or dissimilarities. Treating these data as similarities, would cause the MDS to locate variables with high positive correlations near each other, and variables with strong negative correlations far apart. The distance matrix D, could be the distance between rows (D_{row}), distance between columns (D_{col}) or both. Thus, the output of this MDS will be a set of vectors in p dimensional space such that the matrix of Euclidean distances among them corresponds as closely as possible to some function of the input matrix according to a criterion function called *stress*. These distances could be plotted and will give an idea of how the samples group in a two dimensional space, but they are used to obtain the limits in which the normal operation data lays. This is, find out the minimum and maximum distances for normal operation that will give the exact location of a fault when it is present. Additionally, as the original data matrix could be approximated by the output of the MDS, this output could be taken to look for the distribution that data follows in normal operation. This is done with the help of the percentiles range (PR). This is a robust statistic that will explain the spread of data when they do not follows a normal distribution. In this case, mean and standard deviation are weak statistics and cannot be used. PR shows the variability of the data in the same way the standard deviation does for a normal distributed data set.

A fault could be detected when comparing the PR of a test data set against that of the normal operation. If the PR of the test data is similar to that of the normal operation, then we assume there is no fault present. If a fault is detected in this comparison, a PCA is carried out once again, to look for the variables that are in faulty mode. In order to give a complete diagnosis, it is necessary to locate the period of time or samples numbers when the fault took place. This is done by looking for the position of distances in the output vector of the MDS, which are out of the normal limits obtained in the learning process.

The algorithm for this approach could be summarized as follows:

1. Take a normal operation data set
2. Carry out the feature extraction for learning, this is:
   (a) Perform a principal components analysis PCA
   (b) Split the original data set in smaller windows
   (c) Find a matrix of correlation among variables
   (d) Apply MDS
   (e) Obtain the minimum and maximum distances in order to give the exact location of a fault when present
   (f) Obtain the PR from the output of MDS
   (g) Give a number that describes the data distribution
3. Take a test data set

4. Carry out the feature extraction for testing. Do the same as in step 2
5. Compare the number that describes the testing data distribution against that of normal operation data distribution
6. If data distributions are similar return to step 3 else go to step 7
7. Perform a principal components analysis PCA
8. Look for the variables that have the maximum contribution to the total variability of the system
9. Look for the position of the distances that lay out of limits obtained in step 2 e)
10. Give the diagnosis. Show the variable that is in faulty mode as well as its location

4.4.2. Case Study

This section gives some results to verify the performance of the proposed method. We use measurement data coming from the vehicle simulator VE-DYNA [35]. Six variables were supervised:

1. Left front wheel angular velocity (LFW)
2. Right front wheel angular velocity (RFW)
3. Left rear wheel angular velocity (LRW)
4. Right rear wheel angular velocity (RRW)
5. Longitudinal velocity of the center of gravity (Vx)
6. Yaw velocity (Yaw)

The simulations were carried out over three kinds of vehicle behaviour. These databases are called chicane, speed and slalom.

The normal operation for the V_x variable of the three databases as well as for the yaw variable for slalom database are shown in Figure 23.

Two of the three databases were arbitrary chosen to carry out the feature extraction process to learn from the normal operation mode. The databases

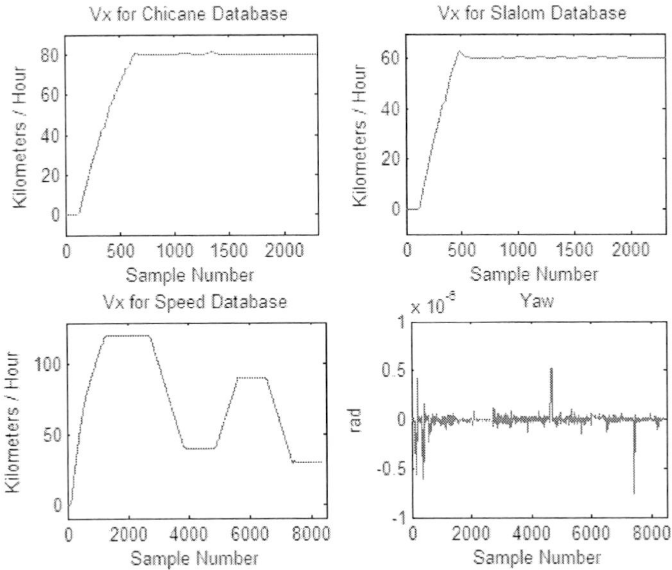

Figure 23. Longitudinal velocity of the center of gravity and Yaw velocity variables for normal operation databases.

selected were chicane and speed. The feature extraction learning block was applied as depicted in figure 24. It shows that PCA was applied to chicane normal operation database in order to determine which of the six variables were correlated and then the same was done to speed database to verify the correlation. To know how many principal components are necessary to retain, we considered the total amount of sample variance explained. We decided to retain the number of PC that explain at least a 95% of the total amount of variance. This analysis gives as a result a division of the six original variables into at most three new variables ($PC1$, $PC2$ and $PC3$). It could be noticed from table 9 that $PC1$ is a weighted average of the first five variables, $PC2$ separates the Yaw velocity from the rest variables as well as it contrasts both left wheels (front and rear) and both right wheels grouped with the longitudinal velocity. $PC3$ groups and contrast both front wheels and logitudinal velocity with both rear wheels.

Table 9 shows the first three PCs to show how PCA groups the correlated variables.

Table 9. Loads of the first three principal components corresponding to the normal operation of the system

Original Variables	$PC1$	$PC2$	$PC3$
LFW	0.4428	-0.3632	-0.4025
RFW	0.3003	0.5121	-0.3500
LRW	0.4290	-0.3923	0.1456
RRW	0.4954	0.2049	0.7742
Vx	0.5320	0.1893	-0.3074
Yaw	-0.0359	0.6143	-0.0214

And in general, these three variables have grouped the four wheel angular velocities and the longitudinal velocity as one variable and it leaves the yaw velocity alone, showing in this way that there is no correlation between them. Thus, for the monitoring of the five correlated velocities a mapping of the normal behaviour was done. This mapping of the entire plot for chicane database was done splitting the total amount of data in windows of 100 samples. The mapping of the data was divided in three zones as depicted in figure 24. These zones were, increasing speed, constant speed and presence of chicane. It was necessary to obtain a normal operation spread of data minimum and maximum interval values for each of the three zones mapped in order to detect when a fault is present. Additionaly it was obtained an interval of minimumn and maximum values for the distances of the samples in order to locate the exact time of the fault. Thus, several windows of 100 samples were taken for each of the three behaviors. Then we calculate the columns correlation matrix for each of them. This is the input of the MDS, which in turn gives as an output a vector of distances from which it were obtained the minimum and maximum values. In this way, when a fault is present, it is possible to compare the distance of faulty samples and locate in which sample or time exactly a fault occurs. This vector of distances was used also to obtain the distribution of data. For the percentiles range we have decided to take the $95th$ and the $5th$ percentiles, such that it gives an interval of values for each mapped zone. Finally the minimum and maximum values of the entire set of intervals given for each zone were taken in order to have the normal operation interval values of spread of data for them. Meanwhile, it was found out that for yaw rate velocity variable it was only necessary

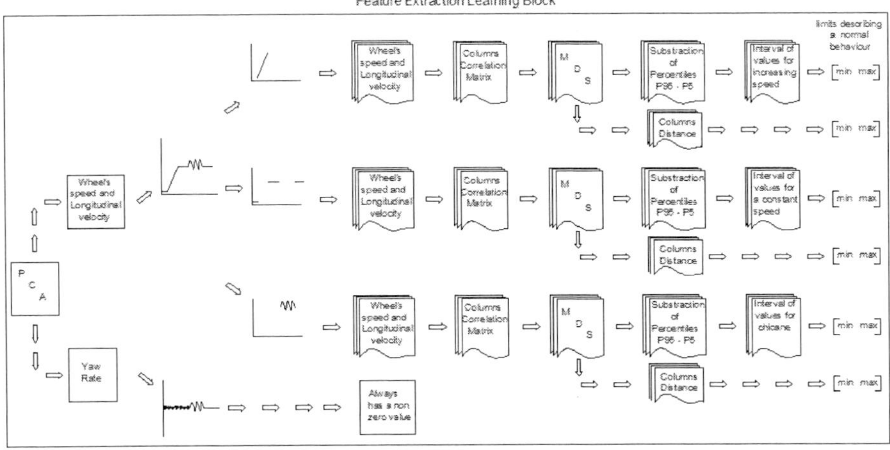

Figure 24. Feature extraction learning block for the simulation example.

to monitor that it always presents a non zero value. Thus, when there is a zero value for this variable a fault is present. When there is a test data set, a very similar procedure is carried out over it. This procedure has a slight difference because as in the learning process, PCA gives two groups of variables here these two variables have been divided since the beginnig and treated separately.

4.4.3. Fault Presence

As the methodology applies the percentiles range over the distances between samples to obtain the spread of data as a measure of the variability on data distribution, and as this PR is obtained subtracting $P95 - P5$ the accuracy of the proposal is measured as follows: When testing a data set coming from a speed constant database it almost always detects the presence of a fault due that the changes present on variability will be registered immediately as when speed is constant there is no variability on data. On the other hand, when testing a data set coming from a chicane or slalom database the percentage of detection will depend on which place lay the faulty data on the test window of 100 samples. If they are located at the beginning or the end of the test window will be different if they are located at the middle of it, additionally this behavior is different if during chicane or slalom the fault occurs at a positive or negative pick or during the slope connecting these two picks. This is due to the variability present on chicane and slalom.

Table 10 shows the percentage of accuracy for the methodology proposed. Percentages in this table were obtained when several faults were induced in a randomly way into different variables and in different locations. As the monitoring process is done taking windows of 100 samples, it was chosen in a random way in which sample number start the monitoring, such that this process lets that the faulty data lay randomly in different positions in the samples windows in order to have different data distributions to test. First column of table 10 shows the percentage of induced deviation in measurement taken, second column is the number of faulty samples induced, detection column concerns to the percentage of success when a fault present, identification column refers to the percentage of identification of the faulty variable and location column is related to the percentage of successful location of the faulty samples.

Table 10. Percentage of accuracy of the methodology proposed when considering a fault of 5% or 3% of deviation present on measurement

Deviation	Faulty Samples	Detection	Identification	Location
5%	15	100%	93.333%	91.667%
5%	10	96.667%	86.667%	90%
5%	5	66.667%	51.667%	56.667%
3%	15	76.667%	65%	71.667%
3%	10	63.333%	55%	61.667%
3%	5	36.667%	21.667%	26.667%

Several simulations have been done to detect simple faults and obtain the efficiency percentage in this task. For instance, a simulation of a single fault present on the left rear wheel angular velocity variable for slalom database is shown in figure 25. A change of 5% was considered on measures during 10 seconds of slalom behaviour.

Table 11 shows the loads of the first three principal components as the total amount of the variance explained by them is 99.9597% for a fault of 5% of measurement deviation from normal operation. It should be noticed that as it is needed to keep three PCs, it is necessary to look for the maximum variability in $PC3$, such that this PC will show the variable that contributes more with the

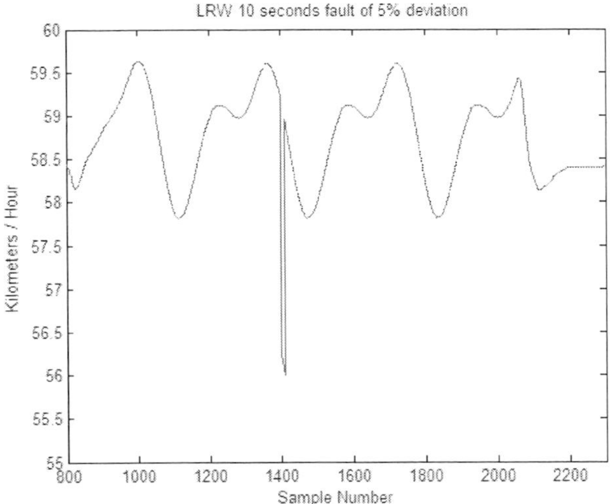

Figure 25. Single fault present on the left rear wheel angular velocity for slalom database.

varibility of the system. In this case this variable is *LRW* as expected. As data was divided in windows of 100 samples and the fault was induced in sample 1400 and the sample window taken was from data 1311 to 1410 the distances obtained after MDS gives that samples from 91 to 100 of the 100 data window taken were the faulty samples as they are the bigger distances, giving in this way the exact location of the fault in this 100 data window.

Figure 26 shows a fault present in the longitudinal velocity variable. This fault is a deviation of 5% during 10 seconds of this variable.

Table 12 shows the loads of the first two principal components as the total amount of the variance explained by them is 99.9597% for a fault of 5% of measurements deviation from normal operation.

As for this case it was needed to keep two PC, *PC*2 has the variable with the major variability and as expected this variable is *Vx*. Once again data was divided in windows of 100 samples and the fault was induced in sample 1800 and the sample window taken was from data 1750 to 1849 the distances obtained after MDS gives that samples from 53 to 61 of the 100 data window taken were the faulty samples as they are the bigger distances, giving in this way the exact location of the fault in this 100 data window.

Table 11. Loads of the PCA for a single fault present on the left rear wheel angular velocity variable having a deviation of 5% during 10 seconds on measures for slalom database

Original Variables	PC 1	PC 2	PC 3
LFW	0.2316	0.6719	-0.6066
RFW	0.4831	-0.0233	0.1690
LRW	-0.1912	0.6983	0.6889
RRW	0.4828	0.0524	0.0367
Vx	0.4784	0.1236	0.0158
Yaw	0.4631	-0.2056	0.3568

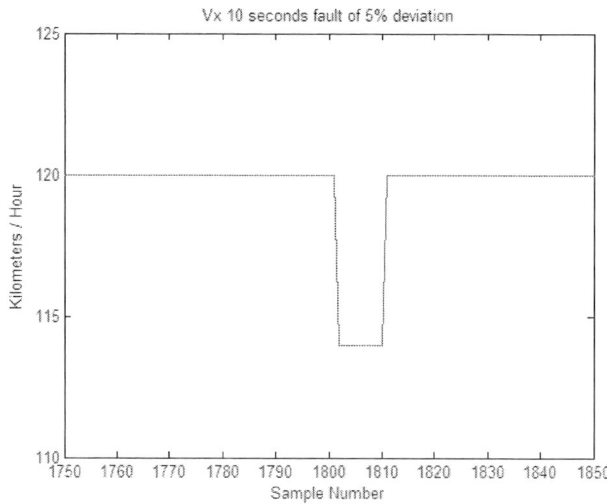

Figure 26. Single fault present on the longitudinal velocity for speed database.

4.4.4. Measurement Noise

After carrying out several simulations it has been found out that the proposed methodology can handle just small magnitudes of noise presence. This is because the method uses the data variability to detect a fault. The results for noise presence are summarized in table 13.

Table 12. Loads of the PCA for a single fault present on the longitudinal velocity variable having a deviation of 5% during 10 seconds on measures for speed database

Original Variables	PC 1	PC 2	PC 3
LFW	-0.7016	0.0	0.0
RFW	-0.7016	0.0	0.0
LRW	-0.0877	0.0	0.0
RRW	-0.0877	0.0	0.0
Vx	0.0	-0.7071	0.7071
Yaw	0.0	-0.7071	-0.7071

Table 13. Results of measurement noise presence

Noise Magnitude	all variables	one variable	two variables
0.00001	100%	85%	65%
0.0001	45%	70%	30%
0.001	0%	40%	0%

4.4.5. Summary of the Framework

This subsection has presented an approach for fault detection based on the combination of process history methods. The methods used were Multidimentsional Scaling, Percentiles Range and Principal Components Analysis in order to carry out a complete system of diagnosis. In order to validate the methodology, the proposal has been implemented for vehicle diagnosis. Several simulations for a six variable vehicle monitoring were carried out and analyzed. This approach could tackle the problem of finding the root cause as it uses PCA and the correlation matrix between variables being monitored in order to find out which is the one that has the major contribution to the variability of the system. Once the detection of a fault has been done, a PCA gives the variable that is in faulty mode and the use of the distances between samples serves to locate the time when it occurs. The proposal could handle just small measurement noise presence due that it uses the distances and variability between samples in order to detect a fault.

4.5. Fault Detection and Diagnosis of a Vehicle Combining an Autoassociative Neural Network and ANFIS

4.5.1. Framework Description

The general fault detection and diagnosis of our last framework is shown in figure 27. As in most of the systems the set of variables are measured in very different scales it is necessary to do a standardization procedure in order to manage all variables over a same scale. Then as depicted in figure 27 the second step is to obtain a subset in a randomly way of the normal operation data composed by the 70% or 80% of the total amount of data, and use these values to perform two things, one is to use them as the learning examples of an autoassociative neural network (AANN) in order to carry out the detection process and the second is to do a correlation analysis between variables.

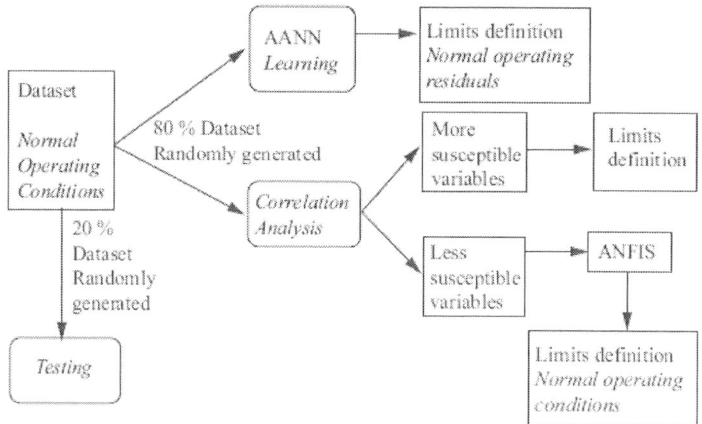

Figure 27. General fault detection and dignosis framework.

As explained in subsection 3.9., the characteristic of a five layer AANN is that it is a net whose outputs are trained to emulate the inputs over an appropriate dynamic range, finding this characteristic to be useful in system wide monitoring. Thus, we implemented it to monitor variables of complex systems that have some degree of correlation with each other as they constitute their inputs. Hence, each output receives some information from almost every input. During training to make each output equal to the corresponding input, the interrelationships between all the input variables and each individual output are embedded in the connection weights of the network. As a result, any specific output, even the corresponding output shows only a small fraction of the input change over a reasonably large range. This characteristic allows the AANN to detect a failure by simply comparing each input with the corresponding output, obtaining in this way the residuals from this comparison. After this, it is necessary to obtain the limits for these residuals of normal operation behavior in order to have the interval in which a variable can range when there is no fault present on it. The correlation analysis will serve to identify which variables are highly correlated given thus the variables where a fault could be reflected even if it is not present on a certain variable. These higher correlations are found by obtaining the correlation matrix of the system and look for the higher correlation coefficients on it (coefficients are positives and negatives). Other way to know if there is a high correlation present between two variables is to obtain the square of the correlation matrix and to look for the correlation coefficients that are higher than 0.3. A third way to know if there is a linear correlation between two variables is to take pairs of variables and plot one versus the other, if there exists a linear correlation the plot will try to be a straight line that could have a slope either positive or negative. If this plot does not forms a line the interpretation is that the variables being related do not have a linear correlation but it could not be said that they are not related as a non-linear relationship could be present. In addition to this, it will be very helpful if an expert on the system is there in order to know which variables are closely related and are seemly hidden or at least it is important to have a brief knowledge of the system in order to look for those relationships between variables. As the AANN will only detect the presence of a fault but not its isolation, it is needed to look for those variables that are more susceptible to small changes on their values, such that when a fault is detected by the AANN, it would be reflected immediately on these variables as the variable in faulty mode will present the higher residuals values and its exact location could be obtained when looking for the samples that lie out of its normal operation limits. If the fault detected by the AANN is not present on the more susceptible variables it is needed to look for it on the less susceptible to small changes variables building in this way an ANFIS as shown in section 3.10. using for this task the corresponding of the formers (that were previously monitored and found in normal operation) as the predictor variables for each one of the correlated rest of variables. Thus, the ANFIS for a specific variable will predict the value of the variable monitored being this its normal operation value. With this value we obtained a normal operation interval, whose upper and lower limits could be taken as ±10% with respect to the normal operation predicted value. Finally a comparison against the normal operation limits is carried out in order to detect which of the variables has the most quantity of points out of them detecting in this way which of the less susceptible to small changes variable is the one with the fault present.

The algorithm is summarized as follows:

1. Take a normal operation data set
2. Randomly take a subset 80% of the total amount of data
3. Normalize the data subset. Train the AANN and learn the model
4. Generate the normal operating conditions residuals
5. Obtain the minimum and maximum value of the residuals (i.e. thresholds)
6. Perform a correlation analysis using either one or all of the following criteria. This allows the identification of variables with the higher or weaker correlations
   (a) Obtain the correlation matrix. Look for the higher correlation coefficients (i.e. the nearest to +1 or -1)
   (b) Obtain the square of the correlation matrix. Look for the correlation coefficients higher than 0.3
   (c) Take pairs of variables and plot them to inspect if the points try to form a straight line (i.e. linear correlation)
7. Build n number of ANFIS relating each of the n variables which are less susceptible to small changes (for instance, those where changes around ±10% are almost unnoticed) in their values, using as predictor variables those in which a little change will be reflected immediately as big residuals at the output of the AANN
8. Obtain the normal operating limits using a ±10% of the values given in 5.

Figure 28 shows the online diagnosis system

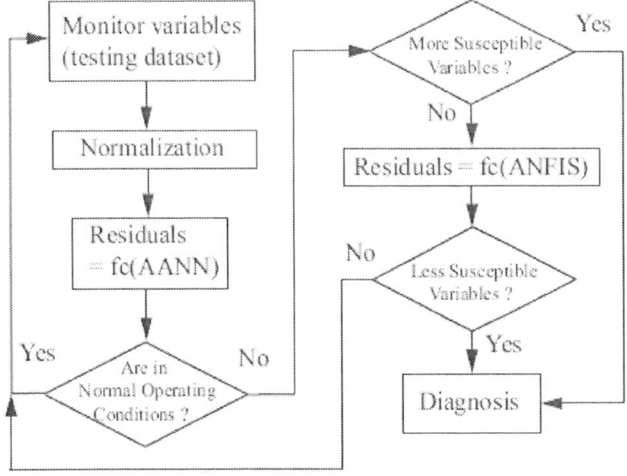

Figure 28. Online diagnosis system.

4.5.2. Case Study

This subsection gives some results to verify the performance of the proposed method. We use measurement data coming from the vehicle simulator VEDYNA [35]. Ten variables were supervised, the first two variables are inputs and the rest eight are the outputs of the system.:

1. Front wheel steer angle (δ_f)
2. Motor couple applied at rear wheels (CMar)
3. Acceleration of the velocity in x (a_x)
4. Acceleration of the velocity in y (a_y)
5. Left front wheel angular velocity (LFW)
6. Right front wheel angular velocity (RFW)
7. Left rear wheel angular velocity (LRW)
8. Right rear wheel angular velocity (RRW)
9. Longitudinal velocity of the center of gravity (Vx)

10. Yaw velocity (Yaw)

The general online diagnosis system can be tailored to this approach as shown in figure 29 and described as follows:

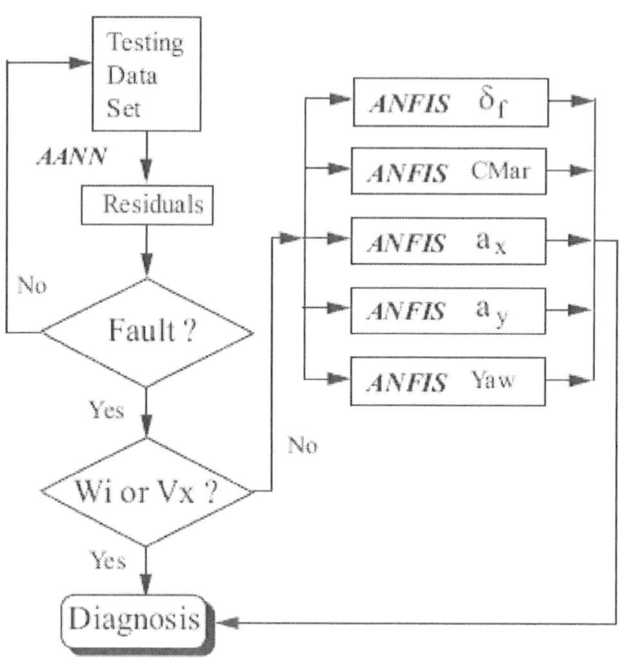

Figure 29. Implementation of the approach in the vehicle monitoring example.

The experiments were performed by simulating a vehicle maneuver in a highway of double change of lane. A database, called chicane was generated from VEDYNA with 501 points, and conformed the normal operating conditions of system. Randomly 400 points (80% of the normal operation database) were obtained of the database and formed a matrix whose dimension is 400×10. This matrix serves as the 400 examples for each of the 10 variables for the learning process of the AANN. We have implemented an [10 15 3 15 10] AANN architecture, that means that it was a 5 layer AANN with 10 neurons on the input and output layers, 15 neurons in the first and third hidden layers and 3 neurons in the bottleneck layer (3 non-linear Principal Components). Figure 30 shows the three non-linear principal components obtained for the normal operation data.

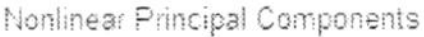

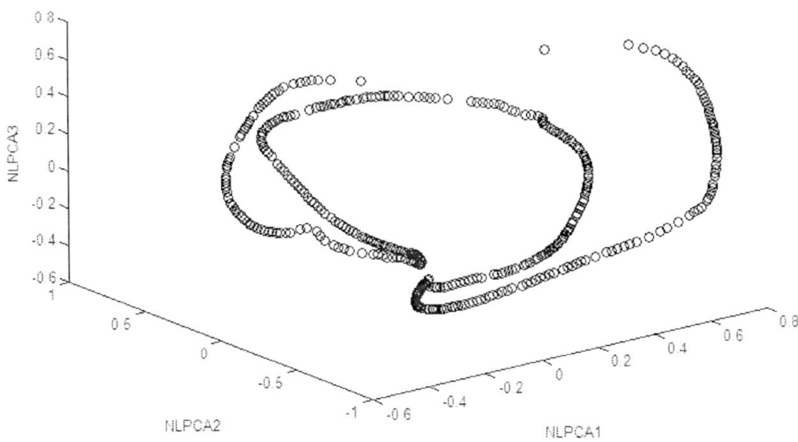

Figure 30. 3 non-linear principal components obtained from the AANN bottleneck layer for chicane normal operation data.

Figure 31 shows the residuals of the AANN and its lower and upper limits for normal operation. Table 14 shows the correlations among the variables based on the correlation analysis.

Table 14. Variables correlated according with the correlation matrix and squared of the correlation matrix criteria

δ_f	a_x	**LFW**	**RFW**	**LRW**	**RRW**	**Yaw**
CMar	a_y	LRW	RRW			

From table 14 it can be notice that the relationship between variables has a lot of common sense. For instance, we noticed that the wheel steering angle (δ_f) is straight forward related with the wheels and that the motor couple applied at rear wheels is obviously related with them. It is necessary to detect which variables will reflect high changes ($\pm 10\%$) in their values at the output of the AANN. This is because the AANN detects a fault based on its variability

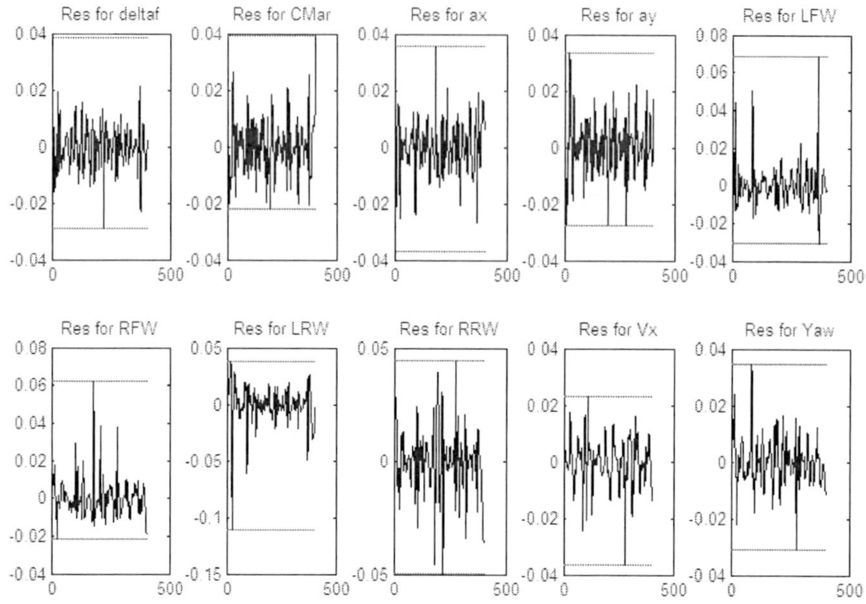

Figure 31. Residuals of the AANN and its corresponding limits for normal operation data.

and it reflects immediately a fault which changes around $\pm 10\%$ of the normal operating conditions. Those variables will be the predictor variables used by the ANFIS. In this case, those variables are the four wheels angular velocities and the longitudinal velocity of the center of gravity (V_x), but the formers were used as the predictor variables for the ANFIS that estimate the values for the rest of the monitored variables (see figure 29).

As rest of the variables were δ_f, *CMar*, a_x, a_y and *Yaw* we implemented five ANFIS each with 81 rules of the form:

- if *LFW* is *A* and *RFW* is *B* and *LRW* is *C* and *RRW* is *D* then *V* is *E*

where *A*, *B*, *C* and *D* are the combination of the membership fuctions low, medium or large and *E* is the predicted value given by the ANFIS or expected value without fault of the *V* variable monitored.

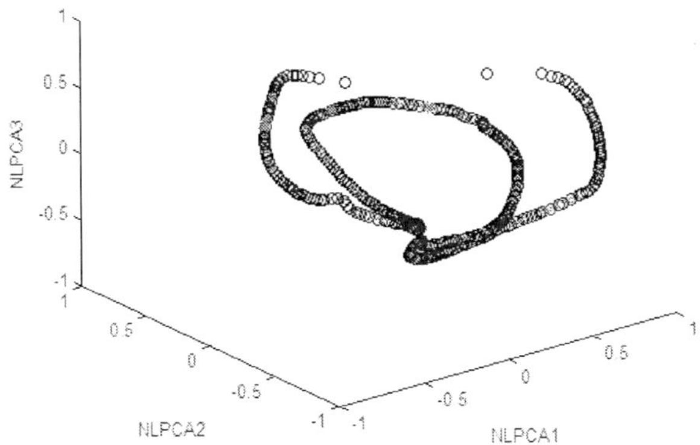

Figure 32. 3 non-linear principal components obtained from the AANN bottle-neck layer for chicane normal operation data used to validate the approach.

4.5.3. Validation of the Approach

To validate the approach we used the rest 20% of the data set of normal operation given by VEDYNA. We proceded as explain above and found the results shown in figures 32 and 33.

Figure 32 shows how the 3 non-linear principal components of the test data set lie on the same space as that of the normal operation database, figure 33 shows how the residuals obtained from the AANN are between their normal operation limits and figure 34 shows how the output of the AANN reconstructs very well its input.

Figure 34 shows the reconstruction of the input variables by the output of the AANN

4.5.4. Fault Present

We carried out several simulations where a fault of different magnitudes as well as in different number of samples was induced in a variable at a time. In the following we show the results of the approach in two examples of diagnosis, the first one when a fault is present in one of the angular velocities of one of the vehicle wheels and the second when a fault is present in the wheel steering

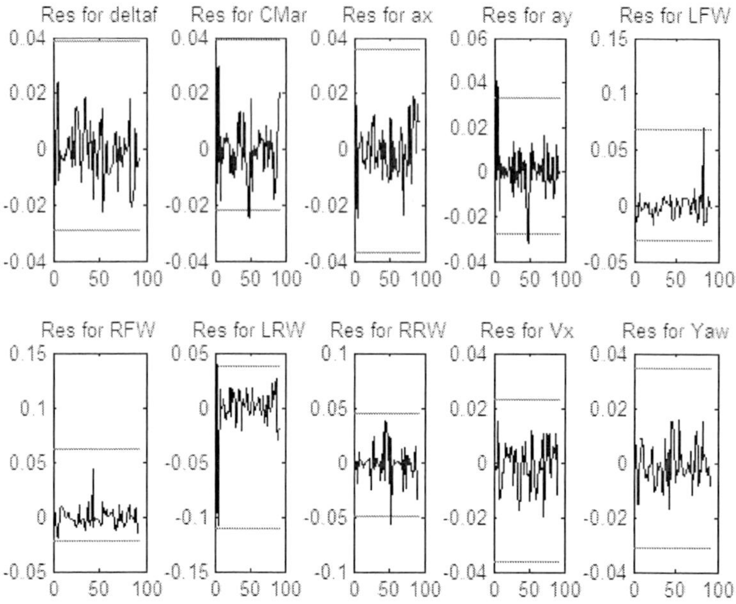

Figure 33. Residuals of the AANN for the validation data set and the corresponding limits for normal operation data previously learned.

angle (δ_f) variable.

4.5.5. Fault Present in LFW

A fault of a change of 11% over the normal operation value in the left front angular wheel velocity (LFW) has been induced from sample number 45 to 57. LFW is one of the variables where a change on their normal values has a big reflection on its corresponding AANN output such that the presence of a fault on this variable is detected in a straight forward way. Figure 35 depicts how the residuals of the AANN show a change on the correlation between the variables of the system and remark (biggest negative residual) this change in LFW due that their normal operation values could reflect a change easily. As this variable reflects a fault immediately at the output of the AANN to give a complete diagnosis all that is needed is to look for the location of the points that are out of their limits for it and as was expected the points that were out of their

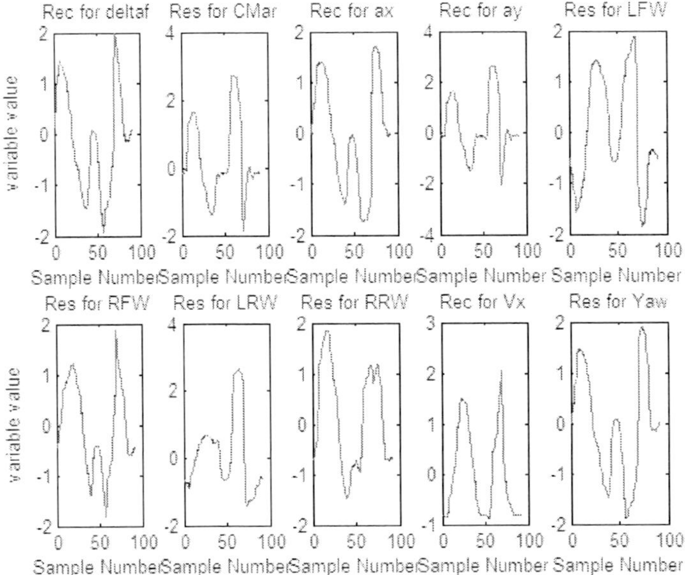

Figure 34. Reconstruction of the input variables by the output of the AANN for the validation data set.

limits were samples from 45 to 57.

Figure 36 shows how the 3 non-linear principal components (NLPC) of the faulty data lie out of the space depicted by the 3 NLPC of the normal operation behavior, indicating thus that a fault is present on the system. Note how it is easy to see the 13 samples that are in faulty mode.

Figure 37 shows how the output of the AANN reconstructs its input reflecting the fault in LFW.

Then the final diagnosis shows that there is a fault present on LFW variable from sample 45 to 57.

4.5.6. Fault Present in δ_f

A fault of a change of 10% under the normal operation value in the wheel steering angle (δ_f) has been induced from sample number 10 to 19. Different from LFW, δ_f is one of the variables where a change on their normal values has a small reflection on its corresponding AANN output such that the presence of a

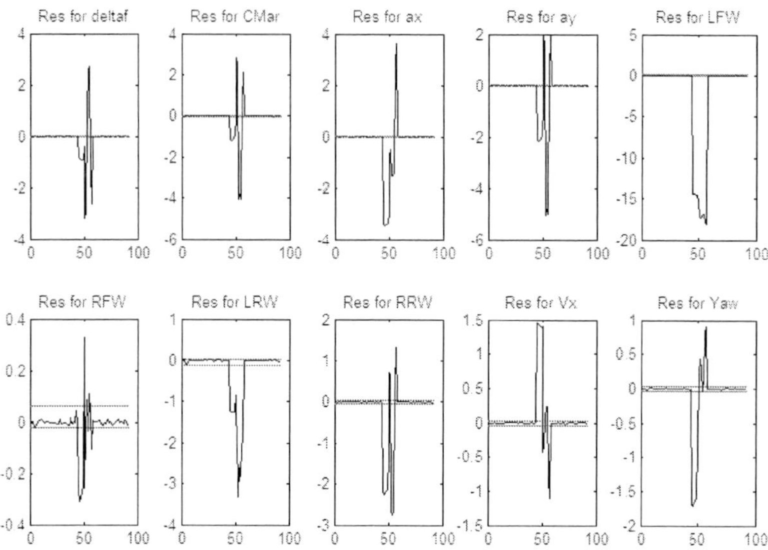

Figure 35. Residuals of the AANN when a fault in LFW was induced over 13 samples.

fault on this variable is not detected in a straight forward way. Figure 38 depicts how the residuals of the AANN show a change on the correlation between the variables of the system and due to the high correlation that this variable has with the others (as this is an input variable for the system being monitored) and to the big variability that it has in normal operation behavior, it is not possible to know which of the variables has the fault on it by simply inspection of the AANN residuals that are out of their normal operation limits.

Figure 38 shows other variables with bigger residuals that those of δ_f that is the real faulty variable.

Figure 39 shows how the 3 non-linear principal components (NLPC) of the faulty data lie out of the space depicted by the 3 NLPC of the normal operation behavior, indicating thus that a fault is present on the system. Note how in this case it was not as easy to see the 10 samples that are in faulty mode as when a fault was induced in LFW.

Figure 40 shows how the output of the AANN reconstructs its input reflecting a fault present in the system, but due to the correlation between variables and

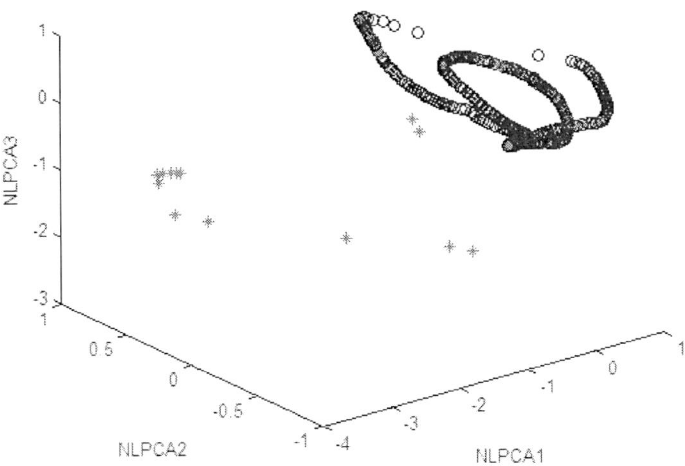

Figure 36. 3 non-linear principal components obtained from the AANN bottle-neck layer for chicane test database when a fault of an increase of 11% in 13 samples of LFW was induced.

due to the big variability in the normal operation values of δ_f it is impossible to know which is the real variable in faulty mode.

Thus, in order to give the diagnosis it is needed to look for the variable that really has the fault on it and the time when it is present. As the residuals coming from the AANN depict a fault present in the system shown that it starts its presence at the same time, but this fault is not reflected on any of the variables whose values could reflect a fault present in a straight forward way (four wheels' angular velocities or Vx) at the output of the AANN, it is necessary to look for the fault in the rest of the five variables. Then, in order to find the real variable in faulty mode of the rest five variables we built an ANFIS for each of them. As described in last section we first search if a fault is present in the variables that could reflect a change in their values and if none of them reflect a big residual (we mean by big residual a change of $\pm 10\%$ in their normal operation values reflected as residuals of magnitudes ≤ -10 or ≥ 10 when plotting the residuals) it is needed to use this variables already checked and found in normal operation

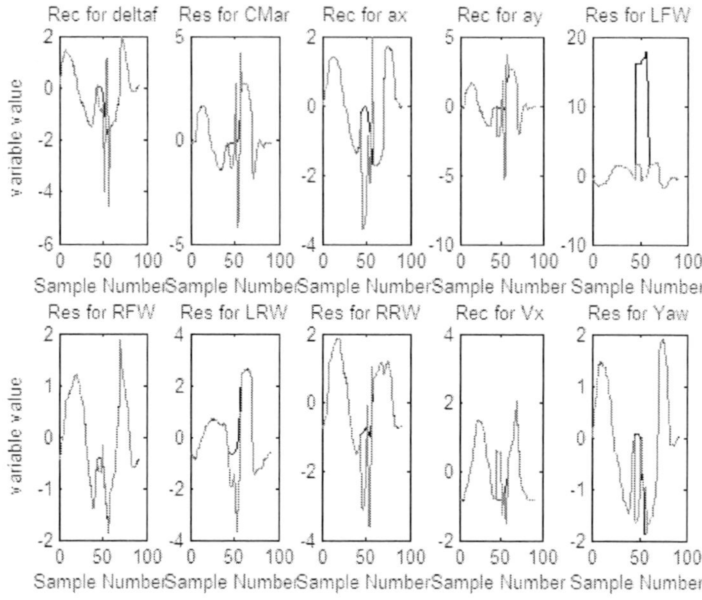

Figure 37. Reconstruction of the input variables by the output of the AANN reflecting a fault in LFW.

as the predictor variables of the ANFIS for those variables that could not reflect big residuals since the output of the AANN (δ_f, *CMar*, a_x, a_y and *Yaw* in our case). Thus, we built five ANFIS using rules similar to the followings:

- if *LFW* is *A* and *RFW* is *B* and *LRW* is *C* and *RRW* is *D* then δ_f is *E*
- if *LFW* is *A* and *RFW* is *B* and *LRW* is *C* and *RRW* is *D* then *CMar* is *E*
- if *LFW* is *A* and *RFW* is *B* and *LRW* is *C* and *RRW* is *D* then a_x is *E*
- if *LFW* is *A* and *RFW* is *B* and *LRW* is *C* and *RRW* is *D* then a_y is *E*
- if *LFW* is *A* and *RFW* is *B* and *LRW* is *C* and *RRW* is *D* then *Yaw* is *E*

where *A*, *B*, *C* and *D* are the membership functions to describe the values of the wheels' angular velocities as low, medium or large, and *E* is the normal operation expected value of each of the five variables monitored.

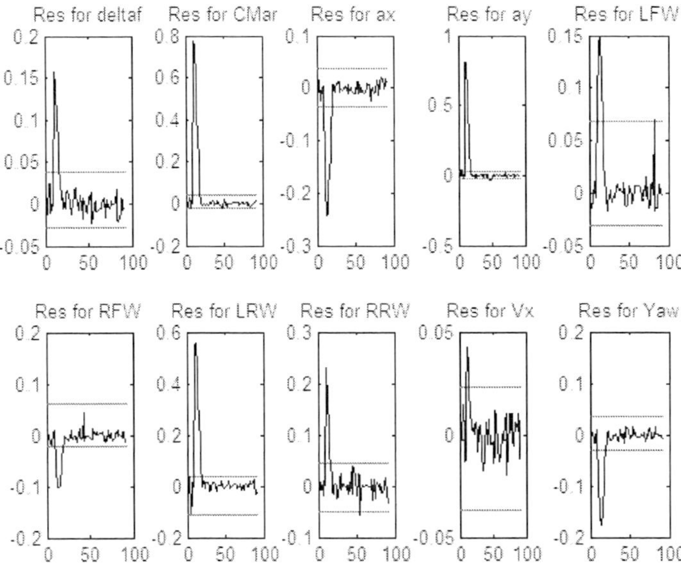

Figure 38. Residuals of the AANN when a fault in δ_f was induced over 10 samples.

Then, these ANFIS are applied to the five variables at the time initially shown by the residuals of the AANN as an initial guess to find the exact location of the fault. Additionally we look around this initial guess of the time when it begins to be sure to find the exact location. Thus, we search 10 samples under and over the initial guess locating in this way in most of the cases the exact samples that are in faulty mode. We specify that it works in most of the cases because in learning process for the ANFIS induce inherent errors on it, such that sometimes the final diagnosis gives a wrong location of the faulty samples. For this example as could be seen on figure 38 the initial guess of the beginning faulty samples was sample number 10 and as we look around this number 10 samples under and over we arrive to locate the exact time where the faulty samples are. So, as expected for this example the final diagnosis gives that the variable that has the faulty data is δ_f and the faulty data are found from sample 10 to 19.

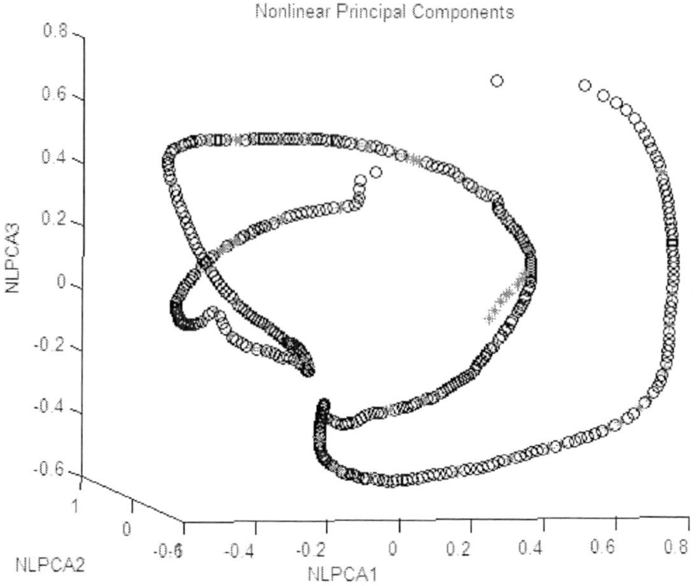

Figure 39. 3 non-linear principal components obtained from the AANN bottleneck layer for chicane test database when a fault of a decrement of 10% in 10 samples of δ_f was induced.

4.5.7. Summary of the Framework

This subsection has presented an approach for fault detection of non-linear systems based on the combination of two process history methods. The methods used were an autoassociative neural network (AANN) and ANFIS in order to carry out a complete system of diagnosis. In order to validate the methodology, the proposal has been implemented for vehicle diagnosis. Several simulations for a ten variable vehicle monitoring were carried out and analyzed. This approach could tackle the problem of false alarms due to the correlation present between variables when finding the root cause as it uses the three non-linear principal components extracted from the bottleneck layer of the AANN. When a fault appears, a first monitoring of the variables more sensitive to small changes on their values is carried out and if no fault is found on them, it starts the evaluation of the ANFIS in order to look for the fault in the rest of the variables less sensitive to small changes in their normal operation values using the formers as

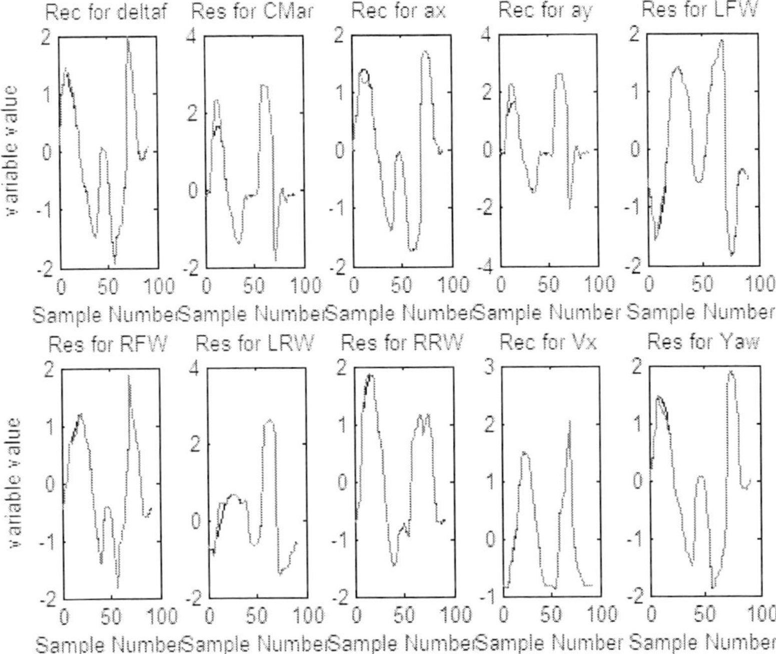

Figure 40. Reconstruction of the input variables by the output of the AANN reflecting a fault in the system.

the predictor variables. As when looking for a fault we compare the samples against normal operation values it is easy to find the location of it.

5. Conclusion

This chapter shows an integration of the frameworks proposed during our research in the fault detection and diagnosis field. It gives an introduction on why it is needed a fault diagnosis system. We showed some research works related to the fault diagnosis field. We also showed how some authors has tackled the problem of complex distributed systems such as electrical power systems and the challenging of non-linear systems such as vehicle systems.

We have given a brief explanation of the mathematical tools that we have used in each of our proposals. We described each of the frameworks proposed and give an application of each in a case study.

It is important to remark that we have decided to implement a fault detection and diagnosis system based only on process history data in order to skip the mathematical model of the system being monitored. This is because in our days almost all the process are automated and is relatively easy to obtain data coming from sensors for the computed controlled systems. It is well known that it is very difficult or even impossible to obtain a mathematical model that could take into account all the variables, noise presence or correlated variables of a very complex system, where it exists all kind of variables correlated or its response could be linear or nonlinear. This is because model based methods depend on designer expertise on the system being modeled and monitored, such that this is reflected in a fault detection and diagnosis system that could easily give false alarms or work with a weak percentage of accuracy.

Another important point is that the only thing that our frameworks need to start is normal operation process history data, which we use as a prior knowledge of the system being monitored. Some of our proposals work with noise presence, giving in this way a good reduction of false alarms. Others of our frameworks can distinguish between noise or fault presence. Another can work with single faults or multiple faults. If there is a fault presence we take into account the correlation between the variables being monitored in order to give as a final diagnosis the root cause of the problem. In general we could say that we have proposed frameworks that are relatively easy to implement and that could work in almost all the industrial processes like, electrical power systems, energy systems, temperature control of a furnace, and are so flexible and powerful that we could implement some of them in vehicle system diagnosis that is one of the most challenging systems due to its non-linerity and variable linear and non-linear correlation. We are sure that the implementation of soft computing methods in fault detection and diagnosis field gives a big and flexible alternative to have a very good and accurate diagnosis system. This is because soft computing methods as we have shown in this chapter works with data coming from sensors and then they use the information directly from system being monitored. We could say that as these methods use process history data they know what is really happening in system, they have a direct communication with the system being monitored because system shows in its history data what it wants us to know about its operating mode. Then we have found that soft computing methods offer a way to communicate in a straight forward way to the system in order to know all about it.

References

[1] V. Venkatasubramanian, R. Rengaswamy, K. Yin, S. Kavuri (2003): A review of process fault detection and diagnosis Part I, Part II and Part III. *Computers and Chemical Engineering* **27**(2003) 293-311.

[2] N. Zanzouri, and M. Tagina (2002): *A Comparative Study of Hybrid System Monitoring Based on Bond Graph and Petri Net Modeling.* 2002 IEEE SMC.

[3] D. Du, X. Lou, and C. Wu (2005): Dynamic Model of FCCU and its Application in a Hybrid Fault Diagnosis System. 2005 *International Conference on Control and Automation* (ICCA 2005), June 27-29, 2005, Budapest, Hungary

[4] W. Wang, X. Bai, W. Zhao, J. Ding and Z. Fang (2005): Hybrid Power System Model and the Method for Fault Diagnosis. 2005 *IEEE/PES Transmission and Distribution Conference & Exhibition: Asia and Pacific Dalian, China*

[5] Q. Peter He, J. Wang, and S. Joe Qin (2004): *A New Fault Diagnosis Method Using Fault Directions in Fisher Discriminant Analysis* . TWMCC Texas-Winsconsin Modeling and Control Consortium. Technical report number TWMCC-2004-05. Department of Chemical Engineering. The University of Texas at Austin, Austin USA.

[6] A. Barigozzi, L. Magni, and R. Scattolini (2004) : A Probabilistic Approach to Fault Diagnosis of Industrial Systems *IEEE Transactions on Control System Technology*, Vol. 12, No. 6, November 2004

[7] W. Liang Cao, B. Wang, L. Ma, J.Zhang and J. Gao (2005) : Fault Diagnosis Approach Based on the Integration of Qualitative Model and Quantitative Knowledge of Signed Directed Graph. *Proceeding on the Fourth International Conference on Machine Learning and Cybernetics* , Guangzhou, 18-21 August 2005

[8] S. Gentil, J. Montmain and C. Combastel (2004) : Combining FDI and AI Approaches Within Causal Model Based Diagnosis *IEEE Transactions on Systems, Man and Cybernetics. Part B: Cybernetics*, Vol. 34, No. 5, October 2004

[9] G. D. Gonzlez, R. Paut, A. Cipriano, D. R. Miranda, G. E. Ceballos (2006) : Fault Detection and Isolation Using Concatenated Wavelet Transform Variances and Discriminant Analysis. *IEEE Transactions on signal processing*, Vol. 54, No. 5, May 2006.

[10] J. Liang, N. Wang (2003) : Fault Detection and Isolation Based on PCA: An Industrial Reheating Furnance Case Study. *International Conference on Systems, Man and Cybernetics*, 2003.

[11] W. Shi, H. Yan, K. Ma (2005) : A New Method of Early Fault Diagnosis Based on Machine Learning. *Proceedings of the Fourth International Conference on Machine Learning and Cybernetics*, Guangahou, 18-21 August 2005.

[12] X. Koutsoukos, F. Zaho, H. Haussecker, and J. Reich (2001): Fault Modeling for Monitoring and Diagnosis of Sensor-Rich Hybrid Systems. *Proceedings of the 40th IEEE Conference on Decision and control*. Orlando, Florida USA, December 2001

[13] Zhang D., Dai S., Zheng Y., Zhang R.,Mu P. Researches and Applications of a Hybrid Fault Diagnosis System P*roceedings of the 3r world Congress on Intelligent Control and Automation 2000*, Hefei, P.R. China, pp. 215-219

[14] Z. Yongli, H. Limin, L. Jinling Bayesian Networks-Based Approach for Power Systems fault Diagnosis. *IEEE Transactions on power Delivery*, Vol. 21, No. 2, April 2006, pp. 634-639

[15] L. Xu, M. Chow Power Distribution Systems Fault Case Identification Using Logistic Regression and Artificial Neural Network *Proceedings of the 13th International Conference on Intelligent Systems Application to Power Systems* 2005, 6-10 Nov. 2005, pp. 163-168

[16] H. Ren, Z. Mi Power System Fault Diagnosis Modeling Techniques based on Encoded Petri Nets. *IEEE Power Engineering Society General Meeting* 2006, 18-22 June 2006

[17] M. Peng, W. Hanli, C. Bin Study of Fault Diagnosis for Power Network based on MAS 2006 *International Conference on Power System Technology*

[18] Bouthiba T., "Fault detection and classification technique in EHV transmission lines based on artificial neural networks", 2005 *European transactions on electrical power*, vol. 15, No. 5, pp. 443-454. 2005.

[19] Hartstein R., Rezende K., and Suman A. "Fault Detection in Primary Distribution Systems using Wavelets", *International Conference on Power Systems Transients (IPST'07)*, in Lyon, France, 2007.

[20] N. Zbiri, A. Rabhi, N.K. M Sirdi, "Analytical Redundancy Techniques for Faults in Vehicle", *Mediterranean Conference on Control and Automation* , Limassol, Cyprus June 27-29, 2005.

[21] M. Schwall, J. Gerdes, "A Probabilistic Approach to Residual Processing for Vehicle Fault Detection", *Proceedings of the American Controls Conference*, 2552-2557, 2002.

[22] M.L. Schwall, J.C. Gerdes, B. Bker, T. Forchert, "A Probabilistic Vehicle Diagnostic System Using Multiple Models", *American Association for Artificial Intelligence*, 2003.

[23] M. Oudghiri, M. Chadli, A. El Hajjaji, "A fuzzy approach for sensor fault-tolerant control of vehicle lateral dynamics", *IEEE International Conference on Control Applications*, Singapore, 1-3 October 2007.

[24] V. Krishnaswami, G. Rizonni, "Model Based Health Monitoring of Vehicle Steering System Using Sliding Mode Observers", *American Control Conference*, Seatle, Washington, June 1995.

[25] P.F. Quet, M. Salman, "Model based Sensor Fault Detection and Isolation for X-By-Wire Vehicles Using a Fuzzy Logic System with Fixed Membership Functions", *American Control Conference*, New York City, July 11-13, 2007.

[26] L. Xu and H. E. Tseng, "Robust Model Based Fault Detection for a Roll Stability Control System", *Transactions on Control Systems Technology*, Vol. 15, No. 3, May 2007.

[27] I. Unger and R. Isermann, "Fault Tolerant Sensors for Vehicle Dynamics Control", *Proceedings of the 2006 American Control Conference*, Minneapolis, Minnesota, USA, June 14-16,2006.

[28] R. Duda, P. Hart, and D. Stork (2001): *Pattern Classification*. Second Edition 2001 ISBN 0-471-05669-3

[29] Russell Stuart, Norvig Peter, 2003 second edition. *Artificial Intelligence a Modern Approach*.

[30] R. Johnson, D. Wichern : *Applied Multivariate Statistical Analysis*. Prentice Hall, fifth edition.

[31] M. Kramer "Nonlinear Principal Component Analysis Using Autoassociative Neural Networks", *Artificial Intelligence in Chemical Engineering Journal*,February 1991.

[32] L. Garza (2001): *Hybrid System Fault Diagnosis with a Probabilistic Logic Reasoning Framework*. Ph.D Thesis, Monterrey, N.L. October 2001

[33] R. Jang, T. Sun, E. Mizutani, *Neuro-Fuzzy and Soft Computing*, Prentice Hall 1997.

[34] H. Tsoukalas, E. Uhrig, *Fuzzy and Neural Approaches in Engineering*, Wiley Interscience 1997.

[35] *Simulator VE-DYNA*. www.tesis.de/index.php.

[36] J.P. Nieto, L.E. Garza and R. Morales (2007) : Fault Detection Combining PCA, Control Charts and Statistic Operation Limits, Research in Computing Science ISSN 1870-4069 *8th International Conference on Computing*, Mexico City, Mexico May 16 to 18, 2007

[37] J.P. Nieto, L.E. Garza and R. Morales (2007) : Multiple Fault Diagnosis in Electrical Power Systems with Probabilistic Neural Networks, IEEE CS Press indexed to EI and ISI proceedings. 6th Mexican *International Conference on Artificial Intelligence Aguascalientes MICAI07*, Mexico Nov. 4-10, 2007

[38] J.P. Nieto, L.E. Garza, A. Rabhi, and A. El Hajjaji (2008): Fault Diagnosis of a Vehicle with Soft Computing Methods. *MICAI 2008: Advances in Artifcial Intelligence*, LNCS 5317:492502, 2008.

[39] J.P. Nieto, L.E. Garza, A. Rabhi, and A. El Hajjaji (2008): Fault Detection and Diagnosis of a Vehicle Combining Multidimensional Scaling. *9th International conference on Sciences and Techniques of Automatic control and computer engineering*. Sousse, Tunisie. Decembrer 20-23, 2008.

[40] J.P. Nieto, L.E. Garza, A. Rabhi, A. El Hajjaji and R. Morales (2009): *Fault Detection and Diagnosis of a Vehicle Combining an Autoassociative Neural Network and ANFIS* Safeprocess 2009, July 2009

[41] J.P. Nieto, L.E. Garza and R. Morales (2010) : Multiple Fault Diagnosis in Electrical Power Systems with Dynamic Load Changes Using Probabilistic Neural Networks. *Computacion y Sistemas magazine*, Vol.14 no.1.

[42] B. H. Lee (2001): Using Bayes Belief Networks in Industrial FMEA Modeling and Analysis. In *Proceedings IEEE of the Annual Reliability and Maintainability* Symposium 2001, pages 7-15

[43] M. Hofbaur, and B. Williams (2004) : Hybrid Estimation of Complex Systems *IEEE Transactions on Systems, Man and Cybernetics. Part B: Cybernetics*, Vol. 34, No. 5, October 2004

[44] M. O. Cordier, P. Dague, F. Levy, J. Montmain, M. Staroswiecki and L. T. Massuyes (2004) : Conflicts Versus Analytical Redundancy Relations: A Comparative Analysis of the Model Based Diagnosis Approach From the Artificial Intelligence and Automatic Control Perspectives *IEEE Transactions on Systems, Man and Cybernetics. Part B: Cybernetics*, Vol. 34, No. 5, October 2004

[45] V. Cocquempot, T. El Mezyani, and M. Staroswieki: *Fault Dection and Isolation for Hybrid Systems Using Structured Parity Residuals* . Labo. d'Automatique, de génie informatique et signal, Ecole Polytechnique de Lille

[46] P. Harihara, K. Kim, and A. Parlos (2003) : Signal-Based versus Model-Based Fault Diagnosis - A Trade-Off in Complexity and Performance SDEMPED *Symposium on Diagnostic for Electric Machines, Power Electronics and Drives*. Atlanta, GA, USA, 24-26 August 2003.

[47] J. Lunze, and J. Schroder (2004) : Sensor and Actuator Fault Diagnosis of System Aith Discrete Inputs and Outputs. *IEEE Transactions on Systems, Man and Cybernetics. Part B: Cybernetics*, Vol. 34, No. 2, April 2004

[48] B. H. Lee (2001): Using Bayes Belief Networks in Industrial FMEA Modeling and Analysis. In *Proceedings IEEE of the Annual Reliability and Maintainability*, Symposium 2001, pages 7-15.

[49] F. Zaho, X. Koutsoukos, H. Haussecker, J. Reich, and P. Cheung (2005): Monitoring and Fault Diagnosis of Hybrid Systems *IEEE Transactions on Systems, Man and Cybernetics. Part B: Cybernetics*, Vol. 35, No. 6, December 2005

[50] D. Poole (1997): The Independent Choice Logic for Modeling Multiple Agents under Uncertainty. *Artificial Intelligence*, 94 pages 7-56.

[51] A.Rabhi, N.K. M. Sirdi and A.Elhajjaji, "Estimation of Contact Forces and Tire Road Friction", *MED'07 The 15th Mediterranean Conference on Control and Automation*, Athens, Greece June 27-29, 2007.

[52] U. Kiencken, L. Nielsen, "Automotive Control Systems" *Springer*, Berlin, 2000.

[53] R. Ramirez-Mendoza. "Sur la modélisation et la commande des véhicules automobiles", *PHD thesis Institut National Polytechnique de Grenoble* , Laboratoire d Automatique de Grenoble 1997.

[54] Pacejka, H. B. "Tyre and vehicle dynamics", *Oxford:Butterworth Heinemann* 2002

In: Fault Detection: Theory, Methods and Systems ISBN: 978-1-61728-291-1
Editor: Léa M. Simon, pp. 177-223

Chapter 4

METHODOLOGIES FOR NOISE AND GROSS ERROR DETECTION USING UNIVARIATE SIGNAL-BASED APPROACHES IN INDUSTRIAL APPLICATIONS

Paolo Mercorelli*
Ostfalia University of Applied Sciences,
Faculty of Automotive Engineering
Robert-Koch-Platz 12, D-38440 Wolfsburg, Germany

This paper addresses Gross Error Detection using uni-variate signal-based approaches and an algorithm for the peak noise level determination in measured signals. Gross Error Detection and Replacement (GEDR) may be carried out as a pre-processing step for various model-based or statistical methods. More specifically, this work presents developed algorithms and results using two uni-variate, signal-based approaches regarding performance, parameterization, commissioning, and on-line applicability. One approach is based on the Median Absolute Deviation (MAD) whereas the other algorithm is based on wavelets. In addition, an algorithm, which was developed for the parameterization of the MAD algorithm, is also utilized to determine an initial variance (or peak noise level) estimate of measured variables for other model-based or statistical methods. The MAD algorithm uses a wavelet approach to set the variance of the

*E-mail address: p.mercorelli@ostfalia.de, mercorelli@dii.unisi.it, Tel: +49-(0)5361-831615, Fax. +49-(0)5361-831602.

noise in order to initialize the algorithm. The findings and accomplishments of this investigation are:

- 1. Both GEDR algorithms, MAD based and wavelet based, show good robustness and sensitivity with respect to one type of Gross Errors (GEs), namely outliers.

- 2. The MAD based GEDR algorithm, however, performs better with respect to both robustness and sensitivity.

- 3. The algorithm developed to detect the peak noise level is accurate for a wide range of S/N ratios in the presence of outliers.

- 4. The two developed algorithms based on wavelets (Algorithms for GEDR and peak noise level estimation) do not require the specification of any additional parameters which makes parameterization fully automated as a result. The MAD algorithm only requires the result of the peak noise level detection algorithm that has been developed and therefore is also fully parameterized.

- 5. The sensitivity and robustness of the algorithms is demonstrated using computer-generated as well as experimental data.

- 6. The algorithms work both, for off and on-line cases where in the on-line case computation times are small so that application on standard Distributed Control Systems (DCS) for large plants should be feasible.

- 7. A translation from Matlab to C (C++) was done using the Matlab Compiler and a COM wrapper was implemented as a functional prototype, which can be directly accessed through an excel spreadsheet.

To conclude, the developed algorithms are totally general and they are present in some industrial software platforms to detect sensor outliers. Furthermore, it is currently integrated in the inferential modelling platform of the unit responsible for Advanced Control and Simulation Solutions within ABB's (Asea Brown Boveri) industry division. Experimental results using sensor measurements of temperature, pressure and Enthalpy in a Distillation Column are presented in the paper.

Keywords: Gross Error Detection; Outliers; Data Protection; Intrusion Detection; Data Reconciliation; Parameter Estimation; Wavelets; Median Absolute Deviation; Variance Determination; Noise Level Determination; Security in Sensor Measures.

1. Introduction and Motivations

Gross error detection and replacement (GEDR) is essential for many model-based or statistical methods in the process control field. An example would be parameter estimation (PE) and data reconciliation (DR). Here, gross error (GE) free data is assumed prior to PE and DR. Currently, only simple GEDR based on limit and rate checks is carried out within some industrial platforms for the detection of outliers. This contribution will use PE and DR as a reference example, however, one should keep in mind that other model-based or statistical methods may also be used instead. As applications in the process industries rely more and more on large amounts of raw data, which either directly originate from a distributed control system (DCS) or historians, there is an increasing need for automated data pre-processing, e.g., cleaning of data. Soft sensoring, data reconciliation, and parameter estimation are example applications [1] that often require the existence of "clean data". In the case of data reconciliation, the data cleaning has been done simultaneously, see [1]. A new method of de-noising which is based on methods already introduced in [2] is presented in [3]. It is based on the comparison of an information theory criterion which is the "description length" of the data. In [4] and [5] this theoretical approach is applied. The description length is calculated for different subspaces of the basis and the method suggests choosing the noise variance and the subspace for which the description length of the data is minimum. In the work, based on [4], it is shown that the de-noising process can be done simultaneously. The approach presented in [4] and [5] is based on "minimum description length" and that's why it seems to be unsuitable for on-line detection because it needs a relatively large amount of data. Furthermore, our approach is looking for a two step de-noising technique. Simultaneous approaches often suffer from the fact that they are based on iterations which in turn impose heavy CPU load requirements on the applications. As a consequence, these approaches often become unfeasible for real-time applications and therefore have not been widely used in industry. Real-time approaches dealing with outlier detection and/or de-noising and featuring an easy means of implementation have also been developed. Historically, lower and upper limits as well as simple rate change check were used to detect outliers. More recently, in [6], [7] and [8] the authors have developed a robust version of the Hampel filter where a lower threshold may be set in order to avoid the undesired detection (removal) of noise. The algorithm assumes knowledge of the noise level of the signal, which is often unknown. The aim of this paper is to develop an approach that addresses outlier detection without de-noising that may be applied in real-time. This is particularly important because some applications require outlier detection without de-noising of the signal. Special attention was given to the fact that the algorithms developed may be tuned automatically during commissioning and may be easily retuned during operation. The strategy consists of using an existing Hampel algorithm (enhanced by [6]) and the newly developed algorithm, based on Wavelets, to determine robustly the noise level of the signal to be processed. The noise level was found to be the parameter of the Hampel algorithm by [6] that cannot be optimally chosen independent of the signal to be processed. The algorithm to estimate the noise is based on Wavelets. The other parameters were optimally fixed using metrics defined by means of simulation studies. Fig. 1 shows a block diagram of how a GEDR package (PCK) would interact with gEST, that is, how the GEDR would function on a conceptual level.

1.1. Modules

The following list describes the modules (blocks) and the connections in order to define the problem statement in general, the key assumptions to be made, and the implications thereof on the GEDR approaches. Dashed blocks and connections denote additional, desired features to be included on a conceptual level, however, they will not be tested or implemented at this point in time. The scope of this work is the GEDR and Quality Module (QM) where the QM has been given more consideration on a conceptual level.

- **Process and Instrumentation**:
  The true process values are affected/changed by sensor (instrumentation) imperfections such as sensor aging or wrong/poor calibration. Often, these effects are modeled using biases. Usually, biases are assumed to be rather small as opposed to Gross Errors (GEs), which may be caused by dead sensors for example.

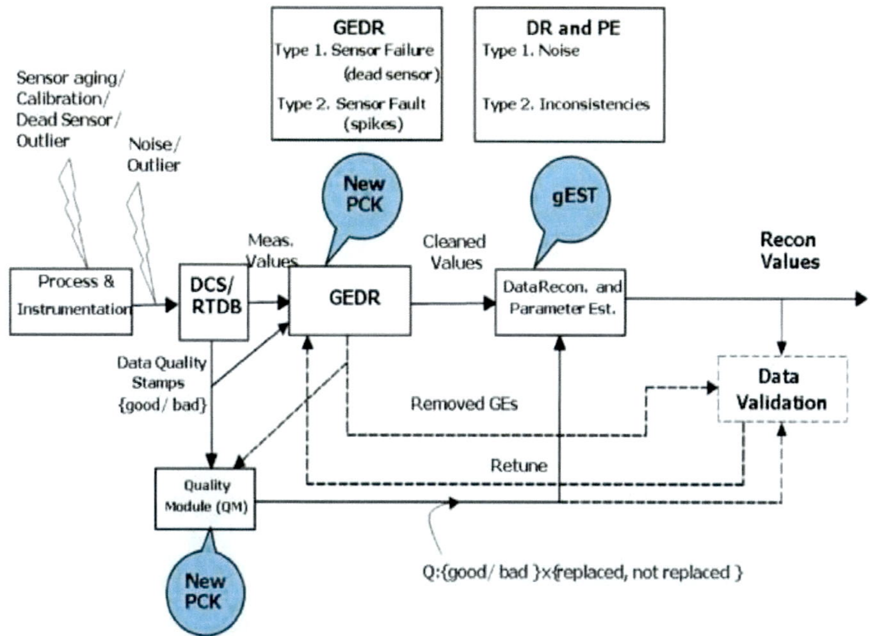

Figure 1. Overview of Modules relevant to GEDR and QM.

- **DCS/RTDB**:
  The DCS/RTDB logs data and stores them in a data base. The data logged consist of the values from the process and instrumentation block and additional effects: noise, outliers, and possibly interference due to the measurement; possible effects to the anti-aliasing; AD-conversion and quantization noise; and possibly LP filtering. Additional information from sensors such as quality stamps may also be available and used within the GED and the Quality Module (QM). For example, a quality stamp may say good or bad.

- **Quality Module (QM)**:
  Some of the functionally of the QM resides within the DCS. For example, in "2 out of 3" set-ups, where the same physical property is measured by three different sensors, the measured value with a "bad" stamp is automatically discarded and values from the other two sensors are used. Different

scenarios, some of which also include soft sensors, are conceivable. The different scenarios strongly depend on the DCS system used and the configuration chosen by the engineers and operators, respectively. For our purposes, the QM module could provide additional functionality using information from the GEDR. Hence, on input the QM merely combines two streams of information. One stream comes from the DCS/RTDB and the other from the GEDR. Characteristics assigned to data points are simply merged and relayed with additional information to other blocks: DR, PE, and Data Validation. The set of characteristics generated by the DCS/RTDB may be good/bad whereas the set generated by the GEDR may be replaced/not replaced.

- **GEDR**:
Gross error detection and replacement must process the data coming from the DCS/RTDB block. The GEDR cleans the data before data reconciliation (DR) and parameter estimation (PE). The GEs, which were detected and removed, should be analyzed after DR and PE to validate that they actually are GEs and in turn to retune the filter settings. GEDR replaces the GE detected with a value that fits the local behavior of the variable. For example, the average of the last couple of the valid data points may be used to do that. This module represents the main part of our investigation.

- **Data Reconciliation and Parameter Estimation (DR and PE)**:
PE is to estimate the parameters whereas DR reduces the error between predicted and measured variables while ensuring consistent process values using redundant information. For example, if all of the mass fractions were measured they would have to add up to one, which represents redundant information. DR and PE must be done after GEDR. This can be done for both dynamic and steady state data. The DR and PE are done simultaneously within some industrial platforms. The DR is performed by introducing BIAS variables that are considered parameters. They are adjusted in order to make the error (difference between measurement and prediction) disappear. Information provided by the QM regarding the quality and possible alternative measurements should be processed within DR and PE. For example, the quality stamp may be: good/badxreplaced/non replaced whereas possible alternative measurements may be a different sensor or soft sensor. Data points identified as being replaced could be weighted differently in order to weaken their influence upon the PE

and DR. Consequently, some provisions regarding the information handling/processing ability of the PE and DR should be made. These may involve, as mentioned, the possibility to change the weights of measured data points in the PE and DR. This is not possible within gEST because the weights for a variable are constant for a given experimental run, that is, no different weights may be assigned for the different data points with respect to time.

1.2. Gross Error Types and Examples

Two types of GEs are considered:

- Type 1 GEs, which are caused by, e.g., sensor failures and result in sustained errors in the measurements.

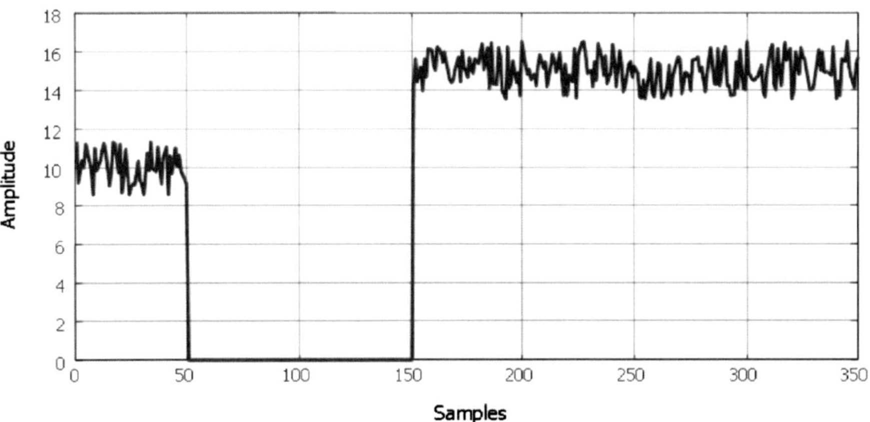

Figure 2. Example of type 1 GE. Computer-generated data with dead sensor defaulting to zero at sample number 50.

Fig. 2 shows an example of a type 1 GE. It is assumed that a sensor goes dead generating a noise-free signal with value 0. The sensor failure occurs at sample number 50 and remains for 100 samples. Provisions for detection of type 1 GEs will be made. However, this will require some user interaction such as specifying the number of constant measurements resulting in a type 1 GE. This option would make sense for variables such as analyzers for which clogging may cause such a behavior. Henceforth, the specification such as the number of constant measurements as well as a dead band should be done by a process expert.

- Type 2 GEs, which are caused by, e.g. sensor faults, result in short, spike-like errors in the measurements

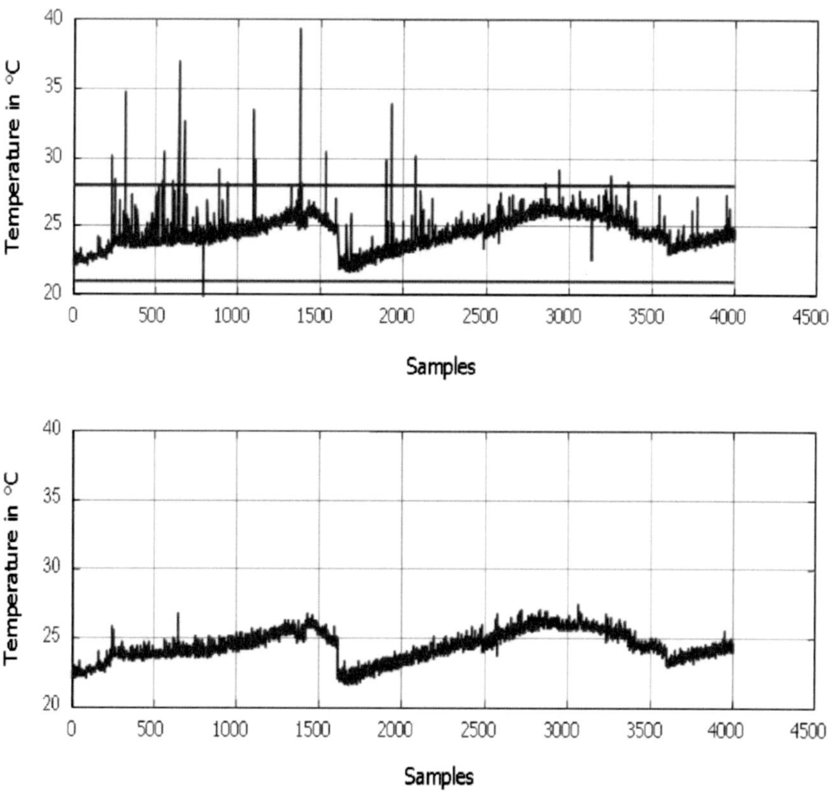

Figure 3. Example of temperature measurements contaminated with outliers (top) and min, max limits. Temperature data cleaned using GEDR (bottom).

Fig. 3 shows data from a temperature measurement contaminated with outliers and hypothetical limits of the min and max values (horizontal thick lines) of the temperature. The limits can be used to perform an easy check to remove some of the outliers. Obviously, not all of the outliers would be removed using the min/max GE check. Another possible easy check could be implemented with a rate limitation. However, this would involve detailed process knowledge of the time constants and noise levels of the measured variables. The GEDR is to remove and replace all of the spikes shown. After detection and replacement, the cleaned data is shown in the lower part of the figure using a MAD based filter, see [6], combined together with the wavelet based peak noise level estimation algorithm. The paper is organized as follows. Section 2. is devoted to the problem specification. Section 3. gives some background elements and presents the state of the art of the noise detection problem. At the end of the section 3. a new wavelet based procedure to detect noise variance is presented and validated. Section 4. shows an algorithm wavelet based for GED and the MAD algorithm with the noise parameter estimated by the new wavelet based procedure. In section 5. and 6. results with computer simulations and real measurements are shown. The conclusions close the paper.

2. Problem Specification

2.1. Mathematical Preliminary

Measurement errors in the context of sampled, measured variables are simply the difference between what is measured and what is the true value of the variable, that is,

$$(y)_k = (x)_k + (o)_k + (b)_k + (w)_k \tag{1}$$

where: $(y)_k$ is measured, $(x)_k$ the true value, $(e)_k = (o + b + w)_k$ is the error due to the measurement and k denotes the $k - th$ value of the time series. In particular o_k is the outlier, w_k is the noise and b_k represents any other kind of fault such as, for instance, a sustained error in the measurements, (see Fig. 2), or systematic errors due to the instrumentation. Equation (1) represents an additive error model as an extension of the outlier model presented in [6, 9]. Another error model used in the literature is a multiplicative error model where the errors are multiplied rather than added up. Often, additive errors are used to model errors occurring at inputs or outputs whereas multiplicative errors are used to model parameter errors, see [6]. Therefore, an additive error model has been chosen. The goal of the algorithms developed is to detect sensor fault related outliers. This results in a short GE that is characterized by a non-continuous, dynamic response such as an impulse or an impulse-like response. Outliers may be distinguished from noise by the amplitude of the peaks. They are short in duration, often no more than two points at a time.

Outlier Detection Problem (ODP) and Algorithm (ODA) Therefore, the outlier index vector has been defined in the following way:

Definition 1 *For a set of data* $(y)_k = (x)_k + (o)_k + (b)_k + (w)_k$ *for* $k = 1, ..., n$ *where* $(y)_k$ *is measured. The number of data points* n *are given. The outlier detection problem (ODP) is to find* $L = (l_1, ...l_n)$ *for which* $(o)_k \neq 0$*, that is:*

$$L = \begin{cases} (l)_k = 1 & (o)_k \neq 0 \\ (l)_k = 0 & elsewhere \end{cases}$$

The index set L itself does not specify what will happen in case an outlier is actually detected. It is supposed to be a mere specification of the detection problem.

The goal of this paper is not to prove some of the inherent features of a data cleaning filter to solve the ODP. This was done in [6]. However, in order to assess the performance of the filter implemented together with the noise detection and to simplify the further discussion some more definitions are necessary. The implementation of an outlier detection algorithm (ODA) will tag the data points within the time series $(y)_k$ that are believed to be outliers. This may be represented by the following definition of a mapping F. If the data point under consideration satisfies the criteria, as used in the Hampel filter in [6], then the map F returns a value equal to one.

Definition 2 *For a set of data* $(y)_k = (x)_k + (o)_k + (b)_k + (w)_k$ *for* $k = 1, ..., n$ *where* $(y)_k$ *is measured. The number of data points,* n*, are given. For an implementation of a filter represented by the mapping* F*, the index vector* L^T *may be computed as follows:*

$$L^T = \begin{cases} (l^T)_k = 1 & (F(y)_k) = 1 \\ (l)_k = 0 & elsewhere \end{cases}$$

The following part will summarize some of the inherent features.

2.2. Noise Level Detection Problem (NLDP) and Algorithm (NLDA)

Definition 3 *For a set of data* $(y)_k = (x)_k + (o)_k + (b)_k + (w)_k$ *for* $k = 1, ..., n$ *where* $(y)_k$ *is measured. The number of data points,* n*, are given. The noise level detection problem (NLDP) is to find the standard deviation for a normally distributed noise* N_l*. Where* $N_l = \|w\|$ *for* $w = (w_1, w_2, ...w_n)$ *and* $\|\cdot\|$ *is some norm. The noise actually detected will be denoted* N_l^T.

The implementation of a noise level detection algorithm (NLDA) calculates a measure of the noise of a time series $(y)_k$. The goal here must be to have an "outlier robust algorithm", that is, the NLDA should not be sensitive to the numbers of outliers present in the time series.

2.3. Some Remarks Regarding Wavelet Based Algorithms

Peak Noise Level and Variance In the context of this investigation, it is referred mostly to the peak noise level estimation. This stems from the fact that the peak noise level estimator reconstructs the peak noise levels of the signal and then calculates their variance. So, when one speaks of a peak noise level estimator, the result will be sometimes close to the standard deviation of the signal and sometimes closer to the actual peak noise level. The goal was to have a robust estimator even for a large range of S/N ratios in the presence of outliers rather than a very accurate estimator for a small range of S/N ratios and no outliers. The proposed technique is wavelet based.

Wavelet Algorithm for GEDR The developed algorithm estimates the local variance of the Lipschitz constant of the signal over a sliding time horizon. The fault (outlier) is recognized if the local Lipschitz constant lies outside the limit computed. Graphically, a flow pipe for the Lipschitz constant is constructed and if the local Lipschitz constant lies outside the flow pipe the data point is flagged and then replaced.

3. Wavelet Based Noise Level Determination

The MAD algorithm uses a wavelet approach to set the variance of the noise in order to initialize the algorithm. In this section this new algorithm, together with the necessary wavelet background, is presented. The developed algorithm estimates the peak noise level. A sliding time horizon is used to estimate a local noise peak. The result for each of the sliding time horizons is stored and used to determine the peak noise level. This is to make the procedure more robust. For an individual time horizon the algorithm selects appropriate wavelet coefficients that represent the noise in the system in the best way.

3.1. Background and State of the Art

The wavelet transform, an extension of the Fourier transform, projects the original signal onto wavelet basis functions, thus providing a mapping from the time domain to the time-frequency plane. The wavelet functions are localized in the time and frequency domain and they are obtained from a single prototype wavelet, the mother function $\varphi(t)$, by dilation and translation. Dilation may be viewed as stretching of the mother function with respect to time whereas translation is a mere shifting with respect to time. Thus, by means of dilation and translation operations performed on the mother wavelet $\varphi(t)$ a set of wavelet functions is created with the same shape as the mother wavelets yet of different sizes and locations. In this work, digital wavelet functions are used to analyze sampled, experimental data. The sets of digital wavelet functions are referred to as wavelet packets. In general, they are characterized by three parameters, two of which may be chosen independently. The parameters depend on the number of the samples considered, that is the data window width, as well as on the Nyquist frequency (half of the sampling frequency). The wavelet packets under consideration are derived from the Haar mother wavelet. Fig. 4 shows a set of Haar functions. The Haar functions are identified using the parameter tuple, (d, j, n). In particular, parameter j is related to the frequency shift, parameter n is related to the time shift, and additional parameter d is useful to address the level of the tree which strictly depends on the number of the considered samples. That is, if eight samples were to be analyzed, the level of the tree would be three. In literature, wavelet functions are structured in the form of a tree such that $d = 1$ represents the highest degree of refinement with respect to time. For example, in Fig. 4 there are four wavelet functions at $d = 1$ whereas there are only two wavelet functions for $d = 2$. Fig. 5 shows the corresponding tree with the wavelet coefficients, $w(d, j, n)$, representing the contribution of each of the wavelets to the signal. The notation $w(1, 0, 0..3)$ denotes the coefficients on the first level on the very left with time shifts 0 through 3.

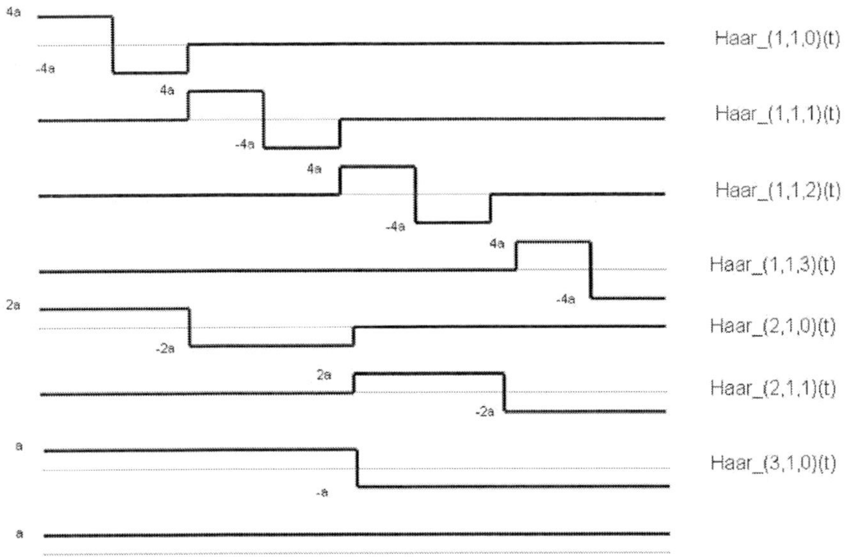

Figure 4. A set of Haar functions.

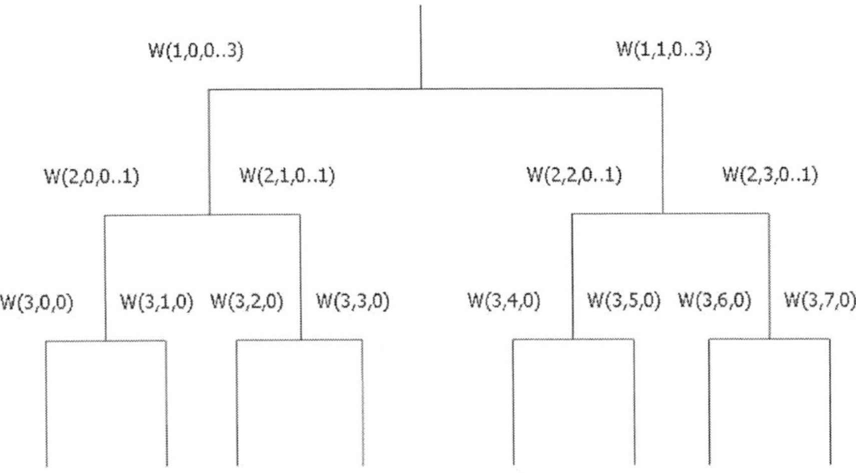

Figure 5. Wavelet coefficients arranged in a tree.

3.2. Noise Level Estimation: State of the Art

Estimation of the noise distribution density is a well-known problem from a theoretical and a practical standpoint. Contributions in this area using wavelets may be found in [10], [9], [11] and [12]. A number of different methods for selecting appropriated threshold values for wavelet de-noising have been proposed. Recent literature [13], with focus on the application domain, addresses the importance of the accurate estimation of noise. In particular, in 3 page 867 a list of threshold criteria is presented. Noise is a phenomenon that affects all frequencies and is present in different forms at different frequencies for different processes. This variability of noise generates difficulty and therefore, heuristic criteria are needed to cope with this. A heuristic used rather often is that the "true" signal tends to dominate the low-frequency band whereas the noise tends to dominate the high-frequency band. The currently used (state of the art) algorithms to estimate noise by using wavelets could be summarized as follows:

- 1 Apply the wavelet transform to a noisy signal and obtain the noise wavelet coefficients.
- 2 Threshold those elements in the wavelet coefficients which are believed to be attributed to noise.
- 3 To reconstruct the noise, apply the inverse wavelet transform to the threshold wavelet coefficients.

In this approach the crucial point is the threshold step, in other words, which wavelet coefficients are believed to be attributed to the noise. In general, wavelet thresholding methods could be classified as soft threshold as in [6], or as hard threshold [14]. Soft thresholds attribute wavelet coefficients, (which are less equal than the soft threshold given), to noise, regardless of the frequencies. Hard thresholds attribute wavelet coefficients, (which are less equal than the hard threshold given), to noise with regard to a minimum frequency given. Both thresholding methods described above may further be sub-divided into two categories: global thresholding and level-dependant thresholding. The two subcategories relate to the level in the wavelet tree. That is, global thresholding applies only one threshold on the wavelet coefficients of one particular level whereas level-dependant thresholding applies level-dependent thresholds on all levels. The latter technique is suitable when the noise is colored or/and correlated. In [10] in 5.4 a practical example of noise estimation is given. The described algorithm uses the finest scale, that is the first level of the tree ($d = 1$, see Fig. 4), of wavelet coefficients to determine the noise. More specifically, the median of the wavelet coefficients at the first level of the tree is used. The use of the median as opposed to the mean makes the estimate more robust because there is a larger number of wavelet coefficients representing noise components of the signal than wavelet coefficients representing "non-noise" components of the signal. The procedure commonly used to detect the noise [10, 9, 11] is as follows:

- **Step 1** Choose the best wavelet basis representation of the measured signal by using the algorithm in [15].
- **Step 2** Decompose the signal using the selected wavelet basis.
- **Step 3** Sort the resulting coefficients and choose the threshold level as proposed in [10] or in an adaptive way on the wavelet tree.

3.3. The Proposed New Procedure for Peak-Noise Level Detection

The proposed procedure calculates the peak-noise level of a sampled signal. To do this, a method is proposed to represent the noise in the wavelet domain, that is, after applying a wavelet transform. The algorithm looks, at every level of the tree, for the time-frequency cell which describes at this level the incoherent part of the signal as defined in [16] and applies a property demonstrated in [16]. The property is in short the following. If the absolute value of the median over n at the time-frequency cell (d, j) (father wavelet) is not more than the absolute of the median at the time-frequency cell $(d + 1, jleft)$ (left child wavelet) plus absolute of the median at the time-frequency cell $(d + 1, jright)$ (right child wavelet) , then the absolute value of the father wavelet is not more than the absolute value of the median of the sum of every couple of left and right children belonging to the same branch of the tree at the level $d + n$ (all the deeper), with d and n integer.

Definition 4 *Given the oscillating part of the sequence $y(t)$ as follows*

$$y(t) - \sum_{n=0}^{2^{N_0-d}} s(d,n)\psi(d,n)(t) = \sum_{j=1}^{2^d-1} \sum_{n=0}^{2^{N_0-d}} w(d,j,n)\psi(d,j,n)(t), \quad (2)$$

where $s_{(d,n)} = w_{(d,0,n)}$. *Then the time-frequency map of the peak values in the wavelet domain is a table of real values specified* $\forall\ d \in \mathsf{N}$ *and by the triplet* $(d, \hat{j}, n)$ *as*

$$w_{P_{(d,\hat{j},n)}} = \max_{j}(\|w_{(d,j,n)}\psi_{(d,j,n)}(t)\|). \tag{3}$$

$\forall\, \hat{j} = 1 : 2^d - 1.$

Remark 1 *The definition intends to select the maximum of the pick values in absolute value at every level of the packet three and at every frequency cell. The idea is to build a map of the pick values.*

Definition 5 *Given an observed sequence*

$$y(t) = x(t) + e(t).$$

Let $e(t)$ *be defined as the incoherent part of the sequence* $y(t)$ *at every level* d *of the packet tree as follows:*

$$e(t) = \sum_{n=0}^{2^{N_0-d}} W_{(d,\hat{j}^*,n)}\psi_{(d,\hat{j}^*,n)}(t),$$

where $\psi_{(d,\hat{j}^*,n)}(t)$ *are the wavelet bases and* $W_{(d,\hat{j}^*,n)}$ *their coefficients. The selected wavelets are characterized by* $(d, \hat{j}^*, n)$ *indices such that:*

$$\{(d, \hat{j}^*, n)\} = arg\Big(\min_{\hat{j}}\big(median_n\{w_{P_{(d,\hat{j},n)}}\}\big) : \\ \{0 < n \leq 2^{N_0-d}, 1 < \hat{j} \leq 2^d - 1, \forall d \in \mathsf{N}\}\Big) \tag{4}$$

where $median_n$ *is the median calculated on the elements which are localized with index* n *(time localized). The coefficients* $w_{P_{(d,\hat{j},n)}}$ *are the wavelet table coefficients of the sequence* $y(t)$ *as given in Def. 4.*

It is easy to see how the idea behind Def. 5 is to sort the basis which can illuminate the difference between the coherent and incoherent part of the sequence, where *incoherent* is the part of the signal that either has no information or its information is *contradictory*. In fact, the procedure looks for the subspace characterized either by small components or by opposite components in the wavelet domain.

Proposition 1 *Let $y(t)$ be a sequence and $w_{p_{(d,j,n)}}$ its corresponding peak values wavelet sequence on different levels of the tree. Because of the wavelet functions are organized in packets, at every level d and at every frequency cell $\hat{j}$*

$$\|w_{p_{(d,\hat{j},n)}}\| \leq \|w_{p_{(d+1,2\hat{j},n)}}\| + \|w_{p_{(d+1,2\hat{j}+1,n)}}\|. \tag{5}$$

Then

$$\|median_n(w_{p_{(d,\hat{j},n)}})\| \leq \|median_n(w_{p_{(d+m,2\hat{j},n)}})\| + \|median_n(w_{p_{(d+m,2\hat{j}+1,n)}})\|. \tag{6}$$

$\forall\, m \geq 1.$

Proof 1 *By observing that, from the orthogonality,*

$$\sum_{(d,j,n)} w_{(d,j,n)}\psi_{(d,j,n)} = \sum_{(d,j,n)} w_{(d+1,2j,n)}\psi_{(d+1,2j,n)} + \sum_{(d,j,n)} w_{(d+1,2j+1,n)}\psi_{(d+1,2j+1,n)}. \tag{7}$$

Functions $\psi_{(d,j,n)}$ are organized in packets and they are scaled functions, $\forall$ d, j and $\forall\, m \geq 1$,

$$\|\max\left(\psi_{(d,j,n)}\right)\| \leq \|max\left(\psi_{(d+m,2j,n)}\right)\| + \|max\left(\psi_{(d+m,2j+1,n)}\right)\| \tag{8}$$

then, from (8) it follows:

$$\|\max_j\left(w_{(d,j,n)}\psi_{(d,j,n)}\right)\| \leq \|\max_j\left(w_{(d+1,2j,n)}\psi_{(d+1,2j,n)}\right)\| + \|\max_j\left(w_{(d+1,2j+1,n)}\psi_{(d+1,2j+1,n)}\right)\|. \tag{9}$$

From Def. 4, $\forall$ d and $\forall\, n \geq 1$, it follows:

$$\|w_{p_{(d,\hat{j},n)}}\| \leq \|w_{p_{(d+1,2\hat{j},n)}}\| + \|w_{p_{(d+1,2\hat{j}+1,n)}}\|. \tag{10}$$

Considering that the "median" is a monotonic function:

$$\|median_n\left(w_{p_{(w,\hat{j},n)}}\right)\| \leq \|median_n\left(w_{p_{(d+1,2\hat{j},n)}}\right)\| + \|median_n\left(w_{p_{(d+1,2\hat{j}+1,n)}}\right)\|, \tag{11}$$

by proceeding $\forall\, m \geq 1$ the proposition is proven.

Remark 2 *In other words the proposition says that, if the minimum of the median at level* d*, according to Def. 4, is not more than the minimum of the median at level* $d+1$ *(one more in depth) then it is also not more than the minimum of the median at level* $d+m$ *(all the deeper), with* d *and* m *integer.*

The introduced definitions and proposition 1 allow the Building of an efficient procedure in order to estimate the level of the noise like an incoherent part of the signal. One can say that the incoherent part of a signal is the part which doesn't "match" the desired signal. The peak-to-peak noise variance is estimated taking the incoherent part of the signal defined above into consideration.

Procedure

Given a sampled signal of length n.

- **Step 0**: Specify length L as $8, 16, 32$, or 64 (dyadic length) and set the end of shifting window $esw = \mathcal{L}$. Determine the height of the wavelet tree: $h = \log_2(\mathcal{L})$. Take the last data points with respect to the end of the shifting window, "esw", of the sampled signal.

- **Step 1**: Construct the wavelet coefficient tree $W(d, j, n)$ for every $d, j > 1$ and for every n, related to the current shifting data window.

- **Step 2**:Build the wavelet coefficients corresponding to the incoherent part of the signal according to Def. 5. In particular for $d = 1$ (the first level of the wavelet tree, this yields $j = 1$).

- **Step 3:**
  Step 3a): For all time-frequency intervals such that $j = 1, 2, ..2^d - 1$ with $d > 1$, calculate the absolute of the median for the same time-frequency interval, that is, $Wmed(d, j) = abs(median_n(W(d, j, \forall n)))$.
  Step 3b): For every "wavelet father" $W(d, j)$ at the node (d, j) with $j > 0$ calculate its left child at the node $(d+1, jLeft)$ and its right child $(d+1, jRight)$. Then calculate the absolute of the medians and denote them as
  $WmedChildLeft = abs(median_n(Wmed(d, jLeft, \forall n)))$
  and
  $WmedChildRight = abs(median_n(Wmed(d, jRight, \forall n)))$. **While** $(d < h)$

For j $=\leq$ $2^d - 1$. **If** $Wmed(d,j) \leq WmedChildLeft + WmedChildRight$, then denote
$Peak(\hat{d},\hat{j}) = \inf_{d,j}(Wmed(d,j))$ and limit the tree of the children of this branch
else
End If
End j loop
End While-d loop

- **Step 4**: Add all of $Peak(\hat{d},\hat{j})$ of the different levels, that is,
$Peak = sum_{(\hat{d},\hat{j})}(Peak(\forall\hat{d},\forall\hat{j}))$.

3.4. Validation of Peak Noise Level Estimation

The two approaches for estimating the variance have been tested.

- 1. Using the de-noising algorithms in the WaveLab toolbox, see [17].
- 2. Using the nlevelmed algorithm as an estimate for the variance.

A summary of the test is that approach 1 is not applicable and that approach 2 is a viable approach. The WPDenoise (m-file) from the WaveLab toolbox has been used, see [17], to get a cleaned signal and then reconstruct the noise by taking the difference between the cleaned and the original signal. However, this is not a sound approach. When the variance of the noise is small, the difference between the cleaned and the original signal will be dominated by the "approximation" error. When there is a high variance then much of the noise will also be present in a cleaned signal. Thus the estimated variance will only be correct for a very short rang of variances. This is illustrated in Fig. 6, where a sinusoidal signal with the added noise has been constructed. This was repeated for a range of variances, and the results showed that only in a small range of variances the estimator gave correct and reasonable values.

For some signals having small S/N it may be impossible to accurately estimate the variance of the noise. Fig. 6 shows an example, a sinusoidal function, where an inaccurate result for small S/N ratios is obtained using a procedure as the one described above.

Using the WPDenoise algorithm from the WaveLab in [17] the results obtained are reported in Fig. 7. The same test is done with the here proposed

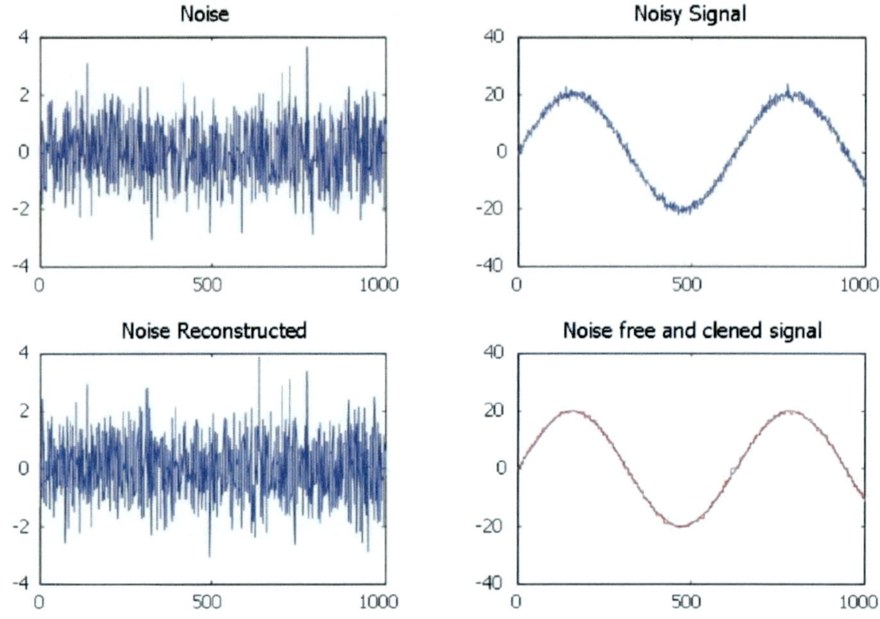

Figure 6. Testing Signal.

procedure and, as it is possible to see in Fig. 8, the results are considerably much better.

Finally, Fig. 9 shows a graphical representation of the distribution of the relative error in determining the noise level error of the data sets. The noise level is determined using a wavelet-based algorithm. It almost follows a normal distribution around $+10\%$, that is, the noise level is underestimated by 10% on average with a standard deviation of 13%. The relative error is defined as:

$$Er = \frac{nLevel - nLevelEst}{nLevel}, \tag{12}$$

where "nLevel" is defined as follows:

- A uniformly distributed noise: "nLevel" = max amplitude of noise.
- A normally distributed noise: $nLevel = 1.5$ of standard deviation of noise.

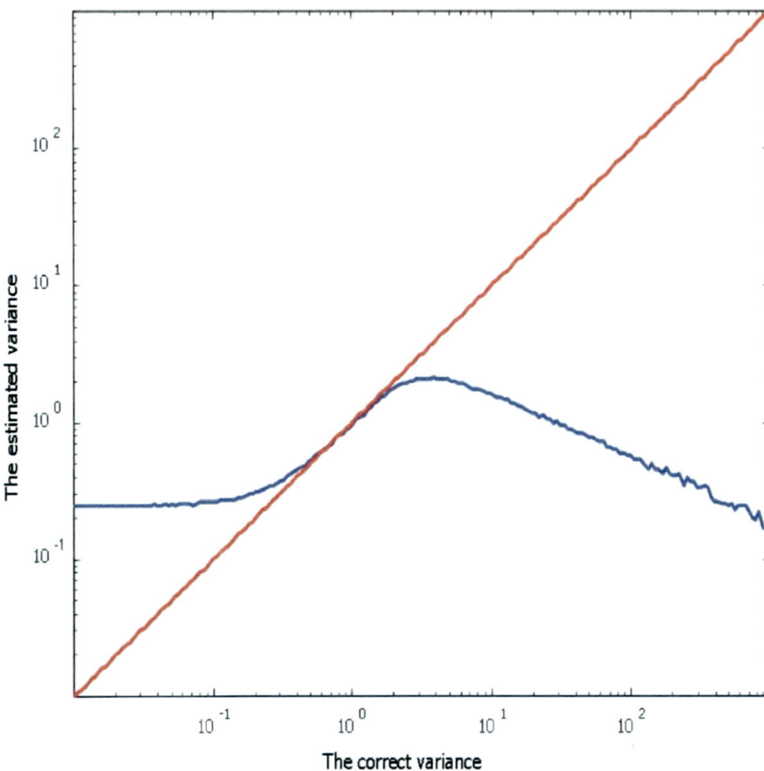

Figure 7. The estimated and true variance using a threshold approach through WPDenoise algorithm from the WaveLab (the worst case).

"nLevelEst" represents the estimated noise level. The noise level detection algorithm was tested for the data sets: dryer, distillation and mining. It does a good job; the noise level always had the right order of magnitude. Fig. 10 shows an example of the noise level for two selected measurements in the data set. It gives an answer which is in the right order of magnitude. See also Fig. 7 and 8 where the estimated noise level is compared with the known variance of a generated noisy signal.

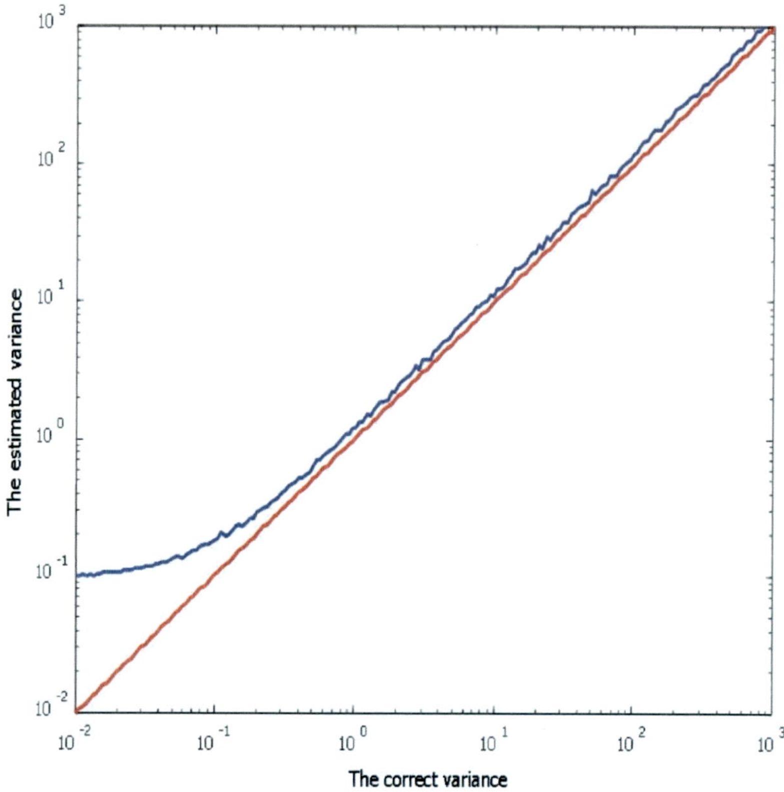

Figure 8. The estimated and true variance using the here proposed procedure.

4. The Wavelet Algorithm for GEDR

The developed algorithm estimates the local variance of the Lipschitz constant of the signal over a sliding time horizon. The fault (outlier) is recognized if the local Lipschitz constant lies outside the computed boundry. Graphically, a flow pipe for the Lipschitz constant is constructed and if the local Lipschitz constant lies outside the flow pipe the data point is flagged and then replaced. One uses a sliding window with length equal to 8 samples in order to compare the results with the procedure proposed in [6]. At each sample time the window is shifted by one sample point. Decompose the signal onto the chosen Haar

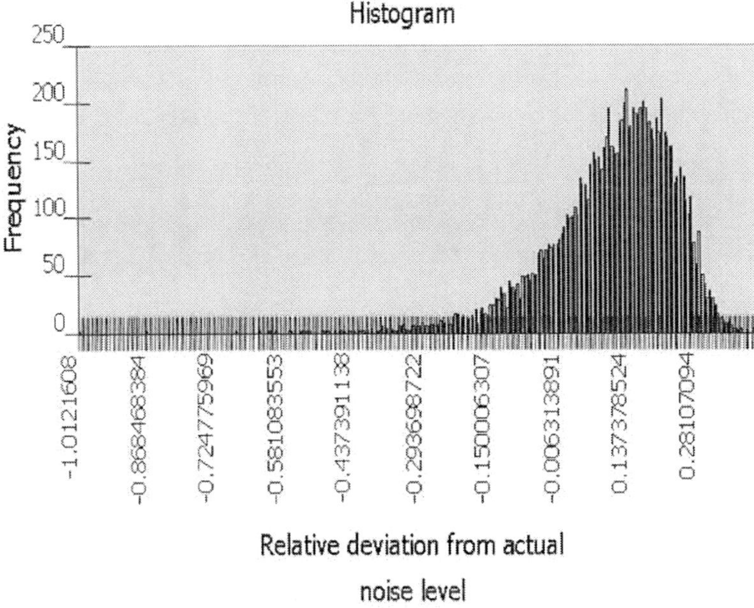

Figure 9. A histogram for Noise Level Detection: the wavelet algorithm for 10000 computer-generated data sets of 200 samples each. X-axis is the relative Error (Er) as defined in equation (12). Y-axis is the density function of real Error with respect to the 10000 computer generated data sets.

function. Start building the function for the first seven samples. With c_1, c_2 and c_3 one indicates the confidential constants.

- **Step 1** Calculate the standard deviation σ of the Lipschitz (L) constant for the first consecutive 7 samples and then one calculates the Lipschitz between the 7th and 8th sample.
- **Step 2** If the last Lipschitz constant is less than c_1 (Where c_1 is equal to 2)[1], then it is not an outlier. Include this Lipschitz constant in the new calculation of σ. If the last Lipschitz constant is greater than c_1, then it is an outlier !!

[1]This value is a heuristic one. The user should test the validity of this value for each kind of signal. Nevertheless, as shown in the here presented examples, the proposed values work for a wide set of signals.

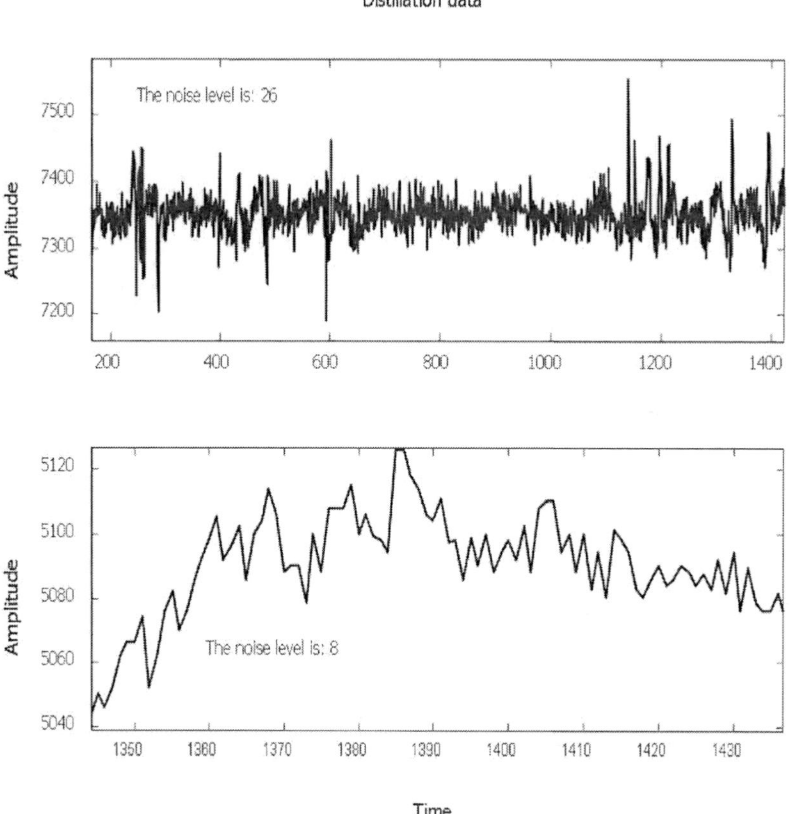

Figure 10. The noise level for two selected measurements which form the distillation data set.

- **Step 3** Store the calculated Lipschitz constant and shift the window in order to consider the next sample of the sequence.
- **Step 4** Calculate the Lipschitz constant between the new $7^{th} - 8^{th}$ sample and add it to the stored Lipschitz.
- **Step 5** If the stored L value is less than $c_1\sigma$, then it is a single outlier and include the last Lipschitz constant in the new calculation of σ, see Fig. 11. If the stored L value is more than c_1, then check if the last two estimated Lipschitz constants have opposite signs, if no then they are multi-outliers,

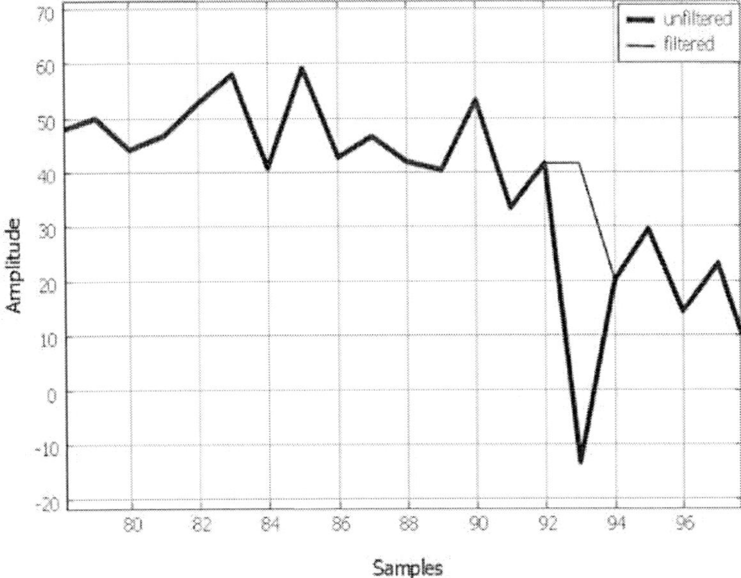

Figure 11. An isolated outlier.

see Fig. 12.
If yes, then check if the stored L value is not less than c_2 (with c_2 equal to 3), if yes, then it is a single inverse multi-outlier, see Fig. 13.
If the stored L value is such that $c_1\sigma \leq L \leq c_2\sigma$, then they are multi-outliers as represented in Fig. 12 or Fig. 14.

- **Step 5a** In case of multi-outliers, shift the window in order to consider the next sample of the sequence and calculate the Lipschitz constant between the new $7^{th} - 8^{th}$ sample and add it to the stored Lipschitz.

- **Step 5b** If the stored L value is less than c_1, then it is not an outlier and include the last Lipschitz in the new calculation of σ. Otherwise check a heuristic safety condition on the local dynamic of the signal performed with another confidential constant c_3, if this is not verified, then they are multioutliers.

With regard to the above mentioned heuristic safety condition one can say that it is related to the dominant dynamic of the signal. In other words, one checks if the dominant dynamic of the signal before the suspected jump is close to the dominant dynamic after the jump. Fig 15 shows an example where this happens.

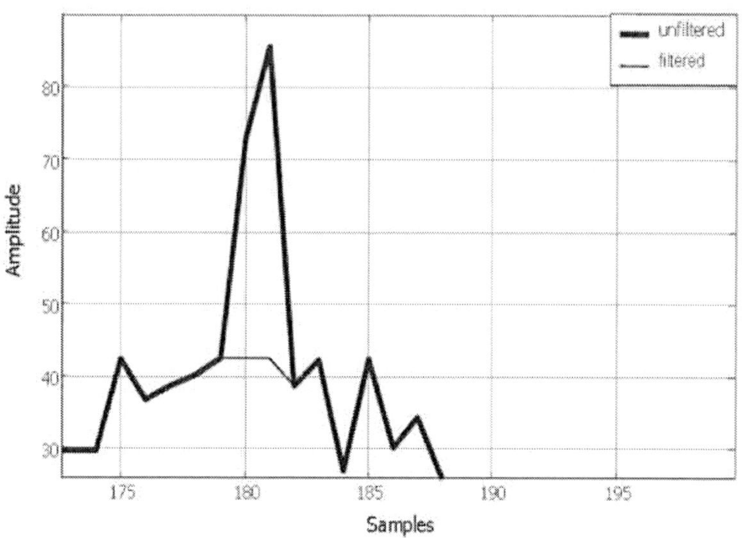

Figure 12. A multiple outlier.

4.1. Validation and Simulations

In this section one validates the procedure by using artificial data where the position of the outliers is previously known.

Fig. 16 is related to the classical approach by using a median filter which needs to know the noise level as a priori knowledge. The results are: 94.68% of outliers correctly detected, 3.6% of outliers incorrectly detected in the data. Fig. 17 shows performance of the wavelet algorithm where one doesn't need prior knowledge of the noise. The results are: almost 100% of outliers correctly detected, 0% of outliers incorrectly detected in the data. It could happen in the proposed procedure that if the outliers are localized on the first part of the data the percentage of the incorrect outliers detected in the data increases and also the percentage of outliers correctly detected decreases. This is due to the initial very small standard deviation of the local Lipschitz constant during the algorithm initialization. This initialization usually takes $15-20$ samples. The weak points

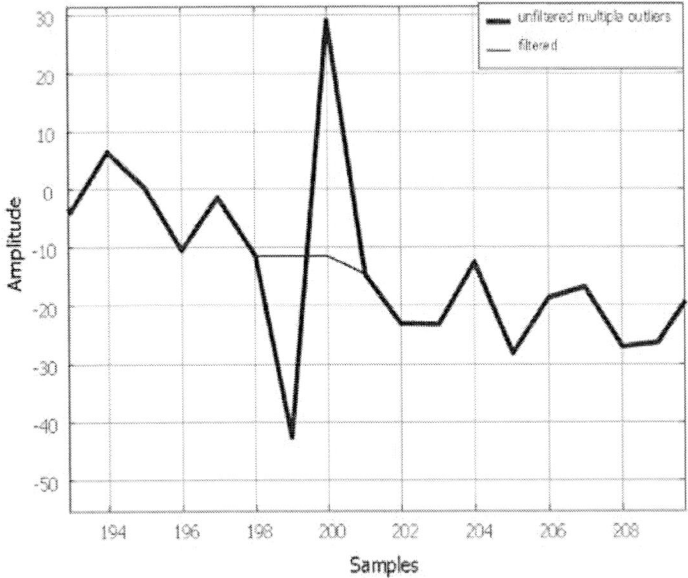

Figure 13. A single inverse outlier.

of the procedure are that it is a stochastic approach and it may not be robust in some cases. For instance, when the above mentioned standard deviation is small, the level of the percentage of the correct detection decreases. In Fig. 17 it is possible to see incorrectly detected outliers during the initialization of the algorithm, that is, in the first 20 samples.

4.2. Outlier Detection Algorithm: MAD Algorithm

The algorithm itself is very simple and is described in [6] in detail. Now the original algorithm [6] and the parameters involved are briefly outlined. Basically, the algorithm uses a moving data window to determine whether or not 1) the center data point, (in the case of an off-line application), 2) or the right-most data point, (in the case of an on-line application) locally fits in the dynamics.

Fig. 18 shows a graphical representation of the MAD approach with the construction of the MAD flow pipe. For our tuning, the parameters were window width=8, thres=4, and var=4*nlevel. Nlevel represents the peak noise level in the data. This was determined to be the optimal setting such that, the flow

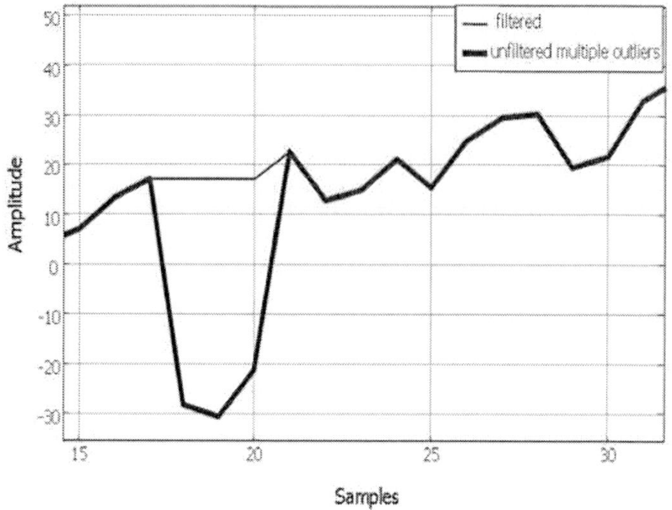

Figure 14. A multiple (3) outlier.

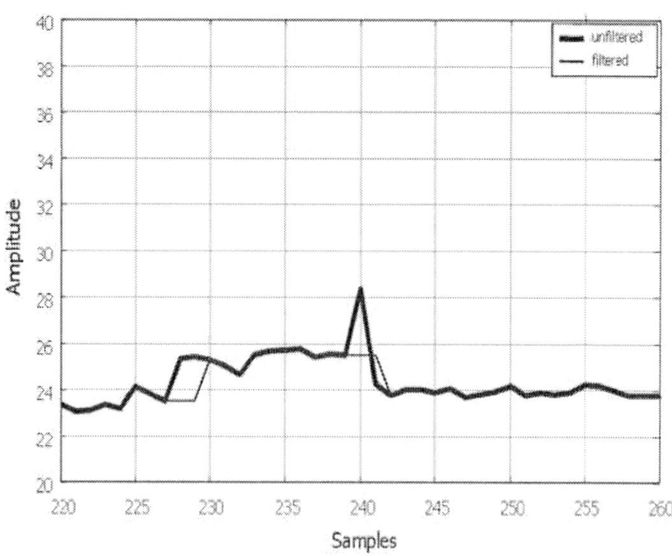

Figure 15. Incorrect detection of multioutliers (on sample number 227 and 228).

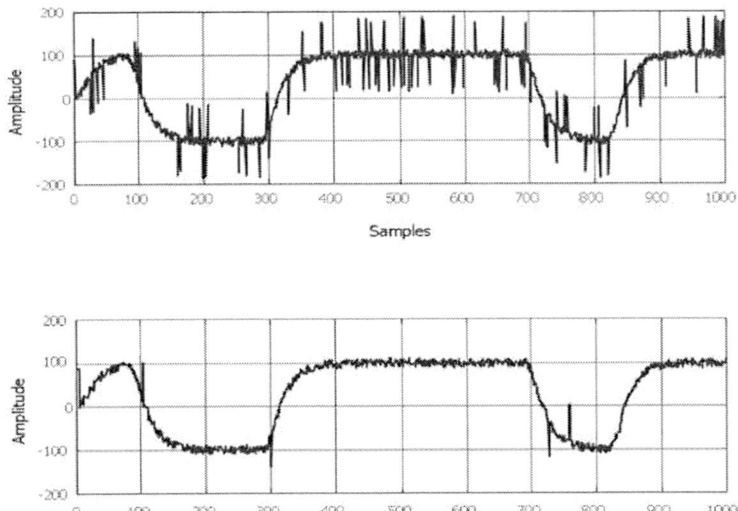

Figure 16. Simulation using median filter with a priori knowledge on the noise.

pipe is wide enough to accommodate the presence of the noise, however, it is narrow enough to remove a high percentage of outliers. Also, it can be seen that the "radius" of the flow pipe is invariant under outliers and noise, which in turn makes it very robust.

5. Results

Assessment Criteria The assessment criteria (metrics) are used to do two things:

- **1.** Find optimal parameters for the MAD and the wavelet GEDR algorithms.
- **2.** Assess the performance of the algorithms using the metrics.
- **1.** Ratio of the detected and removed outliers referred to as ODR (outliers detected ratio). It can be computed both, for experimental and computer-generated data and therefore it will be the only measurement of sensitivity

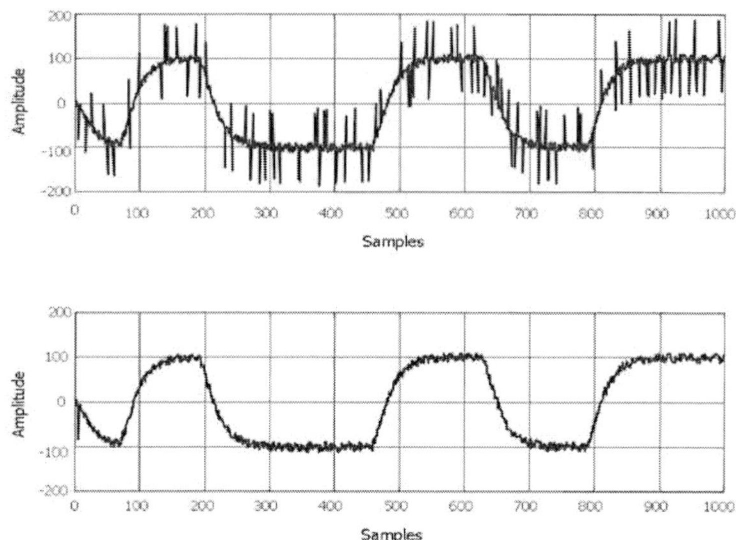

Figure 17. Simulations by using wavelet algorithm without a priori knowledge on the noise.

and robustness for the experimental data because OCDR and OIDR can only be calculated for computer-generated data.

- **2.** Ratio of outliers correctly detected and removed referred to as OCDR (outliers correctly detected ratio). It can be only computed for computer-generated data, where ♯ of outliers which are present in data is known.

- **3.** Ratio of outliers incorrectly detected and removed referred to as OIDR (outliers incorrectly detected ratio). It can be only computed for computer-generated data, where ♯ of outliers which are present in data is known.

- **4.** Introduction of new data using the two-norm over all of the data referred to as SSC (sum of squared changes). It can be computed both for experimental and computer-generated data.

- **5.** Computation Performance (CP). Both algorithms can be computed for MAD and wavelets. Note that the computation is deterministic for both approaches. In the off-line use the computation time is proportional to the

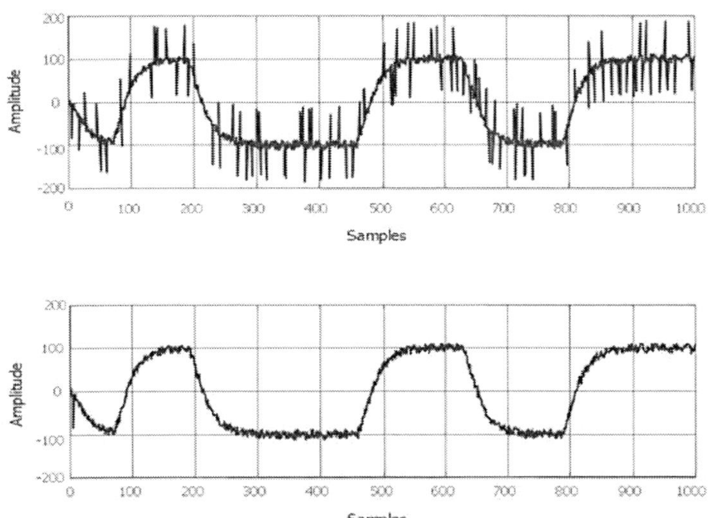

Figure 18. Simulations by using wavelet algorithm without a priori knowledge on the noise.

number of the data points to be filtered whereas on-line it only depends on the window-width of the filter. This metric was not investigated any further because the algorithms are very fast. For example, for the on-line case using the MAD based GEDR the processing within Matlab takes approximately 1ms whereas the processing time for wavelet based GEDR algorithm is approximately 2 ms.

For the presentation of results, only OIDR and OCDR are used to optimize parametric settings for the MAD approach and the tuning of the wavelet-based approach. On the other hand, the CP is not an issue for the developed univariate signal-based approaches. The same applies to the other metrics, which were not found to be useful/critical in this context.

5.1. Algorithm Parameterization

For both wavelet approaches no parameters had to be determined. The algorithms are self-adapting and therefore do not need tuning. The goal was the generation of a reference table such that the optimal settings of the filter for the given signal parameters can be found. An optimal setting of the filter parameters was obtained by using the metrics defined above. The OIDR has been minimized while OCDR at a high level has been maintained, that is, sensitivity for robustness has been compromised. The signal parameters available for the computer-generated data are:

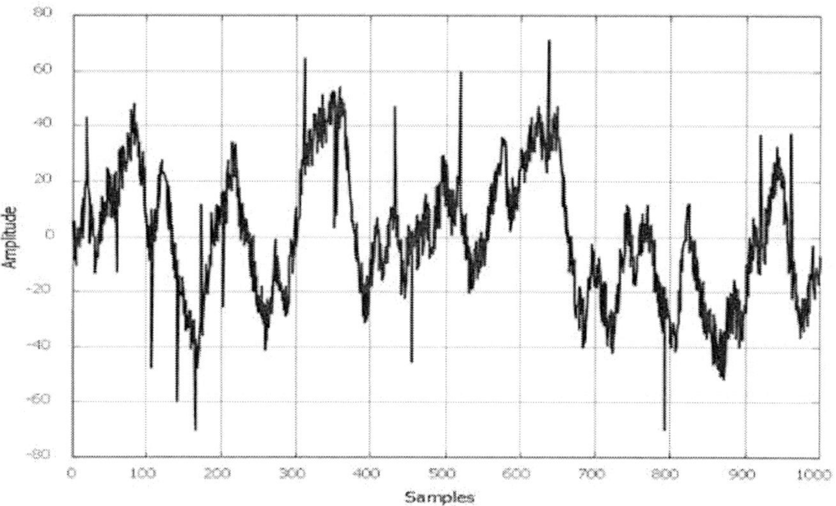

Figure 19. Example of a Computer-Generated Signal: Sampling Time 1, SNR=20, ONR=4,OPR=0.02, DYNF=10, DYNS=20, TRANS=0.1.

- Signal to Noise Ratio, SNR:= (peak to peak signal)/(peak to peak noise) as a dimensionless number. Reasonable SNRs should be between 10 and 100. For a SNR less than 10 the noise with respect to the signal becomes so dominant rendering any kind of meaningful signal processing questionable.

- Outlier to Noise Ratio, ONR:=(peak outlier)/(peak noise) as a dimensionless number. Reasonable ONRs should be greater than 1 (in this application ONRs is used greater than 4) such that they stick out compared to the measurement noise.

- Outlier Probability, OPR: The OPR defines the probability that an outlier occurs at a given time. For all practical purposes, this also represents the ratio of (♯ outliers present)/(♯ data points).

- The fastest time constant ratio, DYNF:=(the fastest time constant)/(tsamp) as a dimensionless number. Reasonable values should be at least 10.

- The slowest time constant ratio, DYNS:=(the slowest time constant)/(tsamp) as a dimensionless number. It should be at least DYNF.

- The number of transients, TRANS:= (the highest frequency of PRBS)/(the sampling frequency). The PRBS is used to generate signals, first order signals with one exponential function, second order signals with two exponential functions, and zero order responses with the PRBS as an input. Reasonable values should be less than 0.5.

- Type: uniformly and normally distributed noise.

Fig. 19 shows an example of computer-generated data with uniformly distributed noise.

The following figures were obtained by randomly generating 10000 data sets of 200 samples each where the above mentioned signal parameters were randomly picked out of a pre-specified range as shown in. Furthermore, the MAD filter parameters, which were optimized as presented above. Again, the wavelet algorithm does not have any parameters. Fig. 20 shows the accumulated distribution of OCDR using the MAD algorithm. The MAD algorithm shows a good level of sensitivity, that is, in less than 10% (y-axis) of the data sets, the algorithm of the OCDR is less than 85%. Fig. 21 shows similar results for the wavelet-based GEDR algorithm. However, for almost 30% of the data sets, the of ODCR is less than 85%. The worst performance is mostly due to the fast dynamics, where 5 was used as a lower limit. As a reminder 10 would be a much better ratio between a sampling rate and fastest time constant.

Fig. 22 and Fig. 23 show a similar representation for the OIDR. Both algorithms show a very good robustness, that is, very few data points were misclassified. However, the wavelet based GEDR performs considerably worse. In case of the MAD based GEDR algorithm the OIDR is mostly about 99.8% of the data sets analyzed in the case of the wavelet based GEDR algorithm.

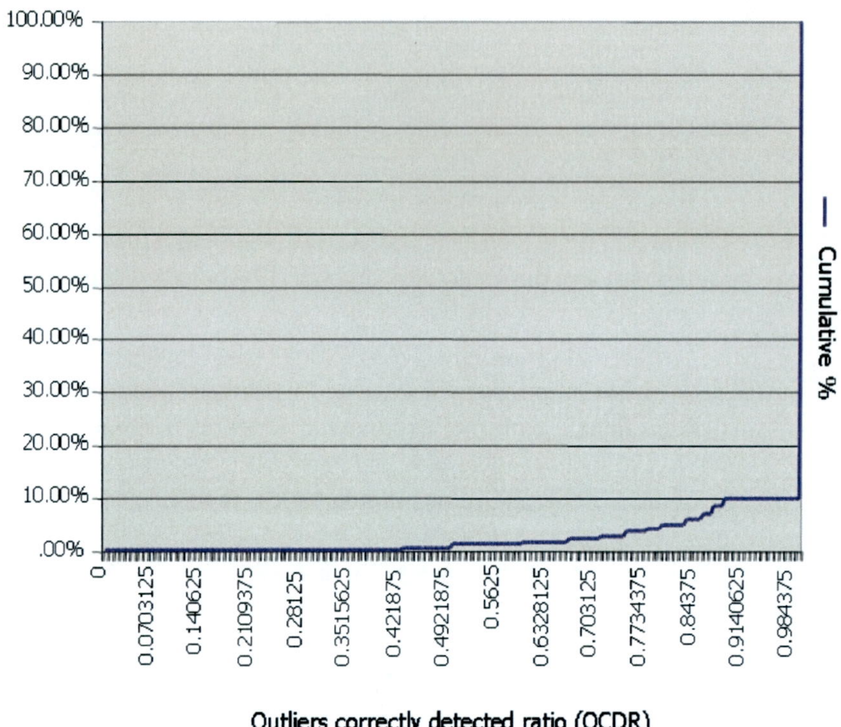

Figure 20. A histogram for OCDR: the MAD algorithm for 10000 computer-generated data sets of 200 samples each. X-Axis shows the OCDR that may range from 0 (no outlier correctly detected) to 1 (all the outliers correctly detected). Y-axis show cumulative % of the data sets for which the "X-property" is true.

6. Experimental Data Sources

Data from different process industries were used to evaluate the developed algorithms. The following list represents the people that provided data with a short description of the process, domain, and the variables involved.

- 1. Separator Train, Oil & Gas : The separator train separates a two-phase flow of gas and liquid. Both phases contain water and oil. The measured quantities are pressures, flow rates, valve openings, and levels, and include process variables, manipulated variables, and disturbances.

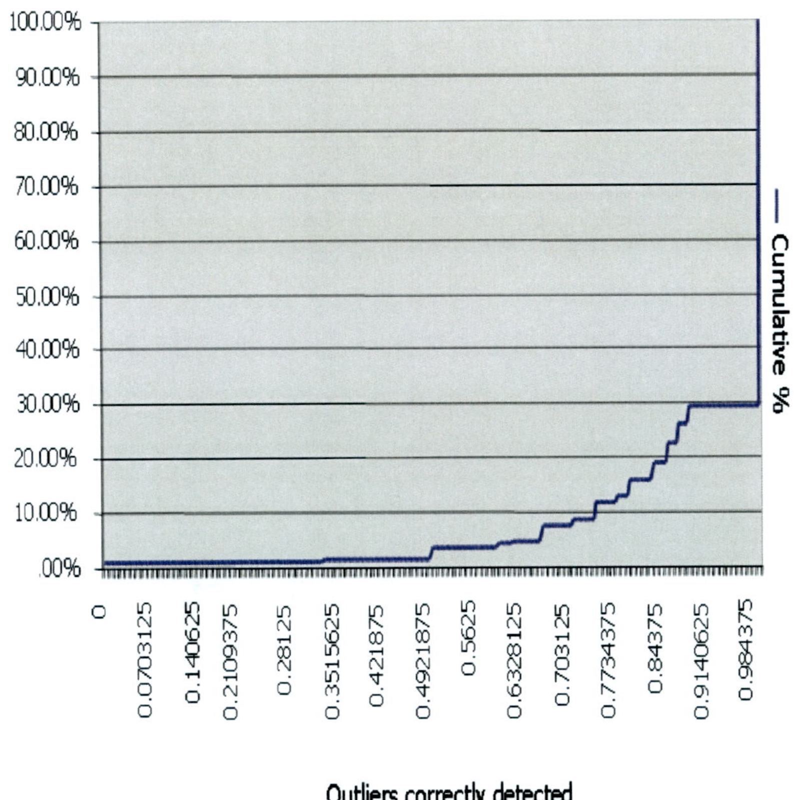

Figure 21. A histogram for the OCDR: the wavelet algorithm for 10000 computer-generated data sets of 200 samples each. X-Axis shows OCDR that may range from 0 (no outlier correctly detected) to 1 (all the outliers detected correctly). Y-axis show cumulative of data sets for which the "X-property" is true.

The variables are strongly coupled and the process shows a strong non-linear behavior. GEs of type 2 are present in the data.

- 2. Dryer Section, Pulp & Paper : This process represents a dryer section within a paper mill and uses steam at different pressures for the drying. The measured quantities are pressures, flow rates, moisture, and levels, and include process variables, set point variables, manipulated variables,

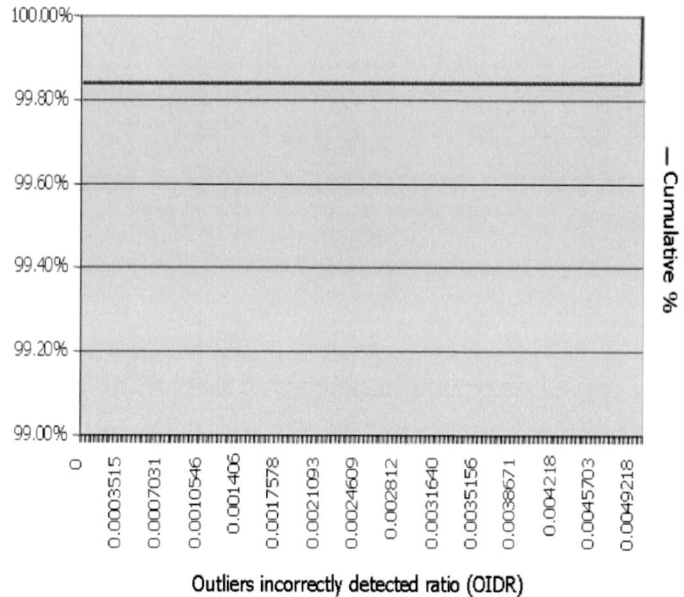

Figure 22. A histogram for the OIDR: the MAD algorithm for 10000 computer-generated data sets of 200 samples each. X-Axis shows OIDR that may range from 0 (no noise misclassified as outliers) to 1 (all noise misclassified as outliers) Y-axis show cumulative % of the data sets for which the "X-property" is true.

and disturbances. The variables are coupled and the process shows a non-linear behavior. GEs of type 2 are present in the data whereas type 1 GEs have not been identified yet.

- 3. Denox, Power Plant This process reduces the amount of NOX in an effluent stream of a boiler. The measured quantities are flow rates and compositions and include a set point variable, a manipulated variable and two disturbances. The variables are coupled, and the process shows a non-linear behavior. No GEs are present in the data due to the "2 out of 3" measurement setup.
- 4. Boiler, Power Plant : This process is a boiler in a power plant. The measured quantities are pressures, temperatures, flow rates, and compo-

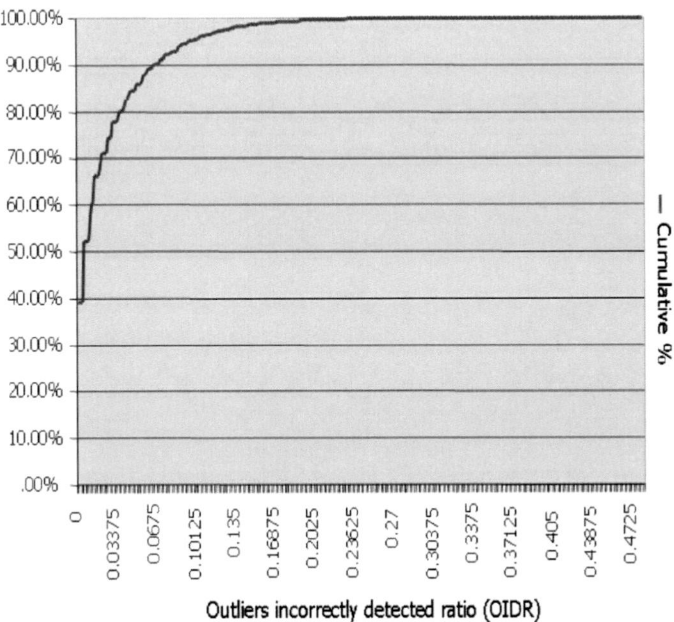

Figure 23. A histogram for the OIDR: the wavelet algorithm for 10000 computer-generated data sets of 200 samples each. X-Axis shows the OIDR that may range from 0 (no noise misclassified as outliers) to 1 (all noise misclassified as outliers) Y-axis shows cumulative % of the data sets for which the "X-property" is true.

sitions. The variables are coupled and the process shows a non-linear behavior. No GEs are present in the data due to the "2 out of 3" measurement set-up.

- 5. Mining Application: The measured quantities are flow rates.
- 6. Distillation Column :The process represents a high-purity distillation column with 13 trays and 4 packed beds. The measured quantities are flow rates, temperatures, pressures and levels. Outliers may be present in the data. No non-linear model is available at the time.
- 7. Discrete Manufacturing Proprietary Information.

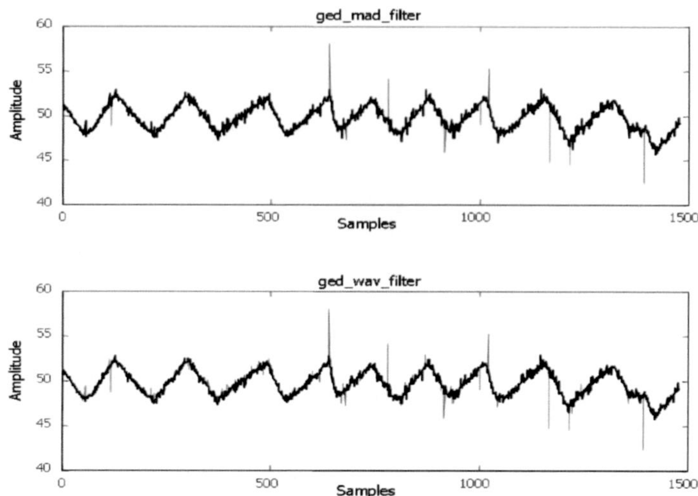

Figure 24. Successful outlier detection in the dryer data set. The fine line is the original data set and the bold is the filtered one.

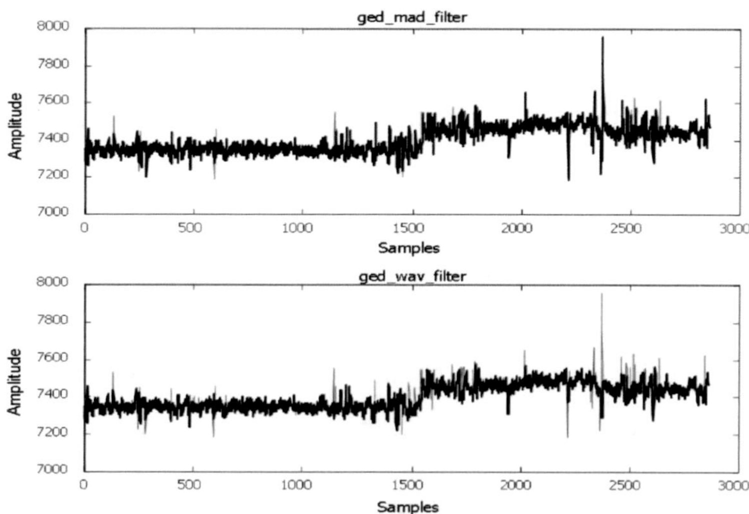

Figure 25. Outlier detection. The fine line is the original data set and the bold is the filtered one.

6.1. Dryer, Distillation and Mining Data with Outliers

Only the dryer data of the examined data sets were clearly contaminated outliers. There were outliers in pressure measurements, which were successfully identified both by the MAD and the wavelet filter. As it is shown in the upper plot in Fig. 24, the "ged_mad_filter" (MAD filter) removes the most obvious of the outliers, although there are probably some more of them. However, it does not remove any measurement noise. The "ged_wav_filter" (wavelet filter) removes more of the outliers, but it also removes some points that are clearly measurement noise.

Fig. 25 shows an example from the distillation where it is uncertain if one is removing outliers or measurement noise. Again it is possible to see that the wavelet filter is more aggressive than the MAD one.

Noise Classified as Outliers As Fig. 26 shows there are cases were it is uncertain if it is a noise measurement or if it is an outlier. Fig. 26 shows an example where both of the algorithms remove noise in an outlier free data set.

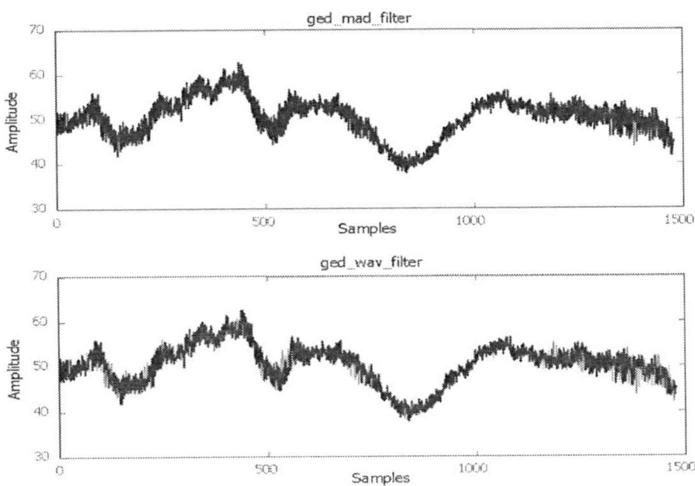

Figure 26. Removal of noise. The fine line is the original data set and the bold is the filtered one.

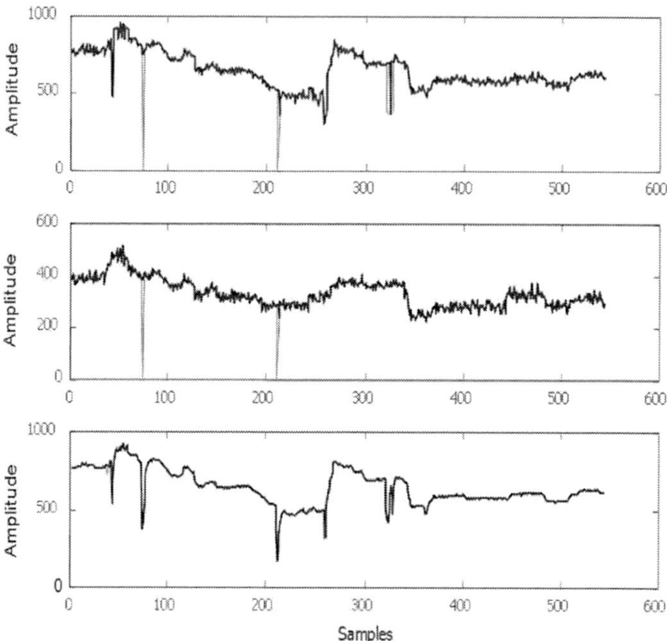

Figure 27. The "ged_mad_filter" applied to the Mining data. The fine line is the original data, and the bold is the filtered one.

Correlated Data and Outliers Since the given algorithms are univariate such effects will not be taken into consideration. In Fig. 27 and Fig. 28 the original and the filtered data are shown. Figure shows that some outliers are detected, but only in two of the three measurements. However since this peak is present in all three of the variables and that they are strongly correlated, it is unlikely that it is an outlier. In Fig. 28 the same results are shown for the wavelet filter, where at least the spikes are removed in all three of the measurements.

Misclassifications For the gross error detection algorithms to work properly it is required that the sampling period is one tenth of the "dominating" dynamics of the process. Below, an example has been shown in which the algorithms have some problems. This may be explained in the following way: the sample time chosen is too long when compared with the dynamic of the signal. Therefore, in Fig. 29 the "ged_mad_filter" has problems due to the oscillating data.

Fig. 30 and Fig. 31 show a problem with the wavelet approach, the signal changes are probably too fast for the filter to keep up. The stair case step is introduced due to the way the replacement is done.

6.2. Artificially Contaminated Data and Off-line, On-line Mode

Boiler Data and Off-line, On-line Mode The boiler and the denox data are both outlier free. The data were used to show the effect of running the MAD based GEDR algorithm off-line and on-line. Again, the wavelet based peak level noise estimation algorithm was used to parameterize the MAD based GEDR algorithm.

Fig. 32 shows the result obtained from the off-line mode. The top part shows raw data from the boiler where no outliers are present. The algorithm does not detect or replace any outliers, that is, the algorithm does the right thing. The

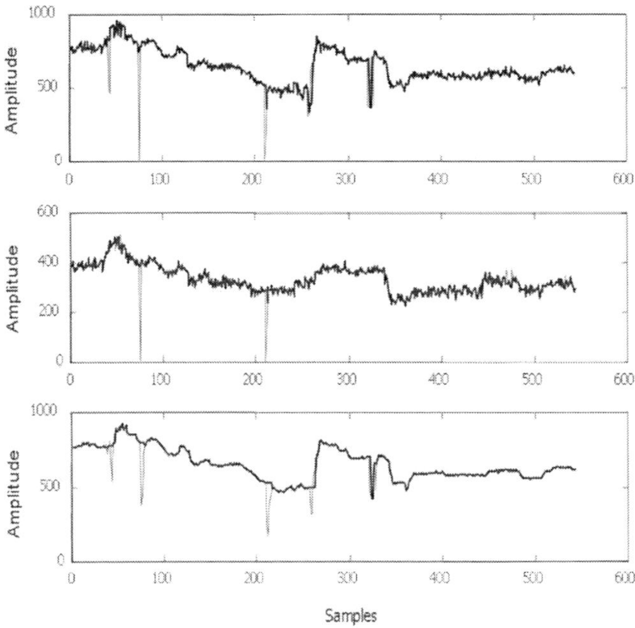

Figure 28. The "ged_wav_filter" applied to the Mining data. The fine line is the original data, and the bold is the filtered one.

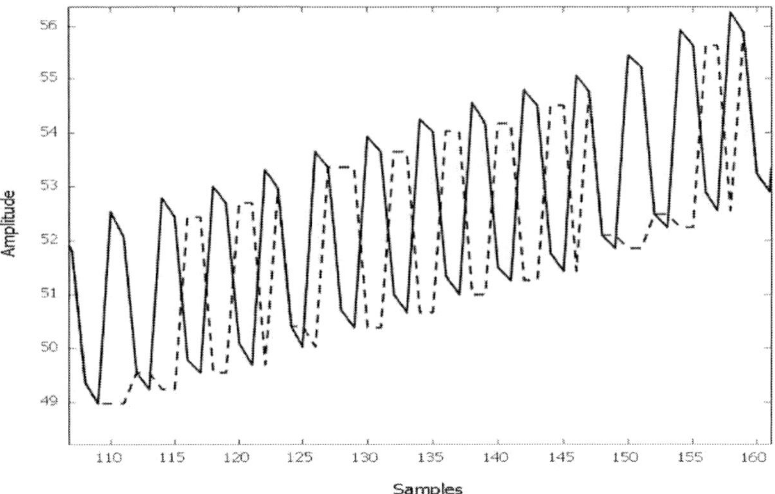

Figure 29. Problems when the dynamics are near the step size for the "ged_mad_filter". The fine line is the unfiltered and the bold is the filtered one.

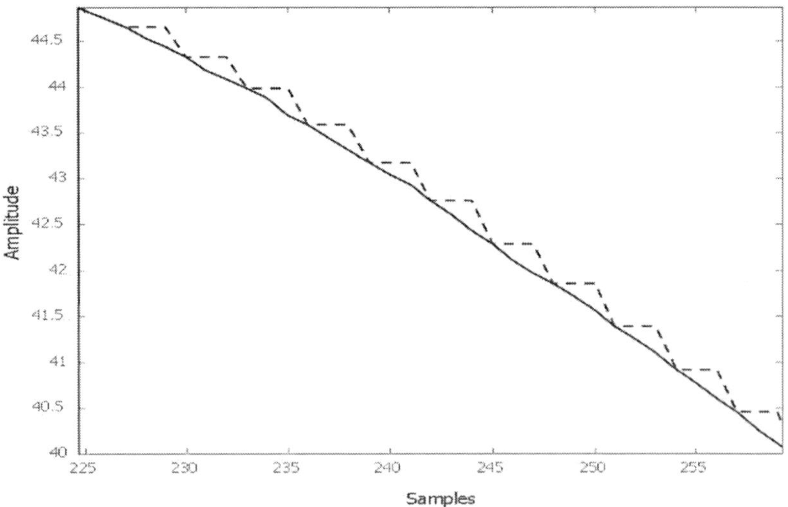

Figure 30. Problems for the wavelet approach.

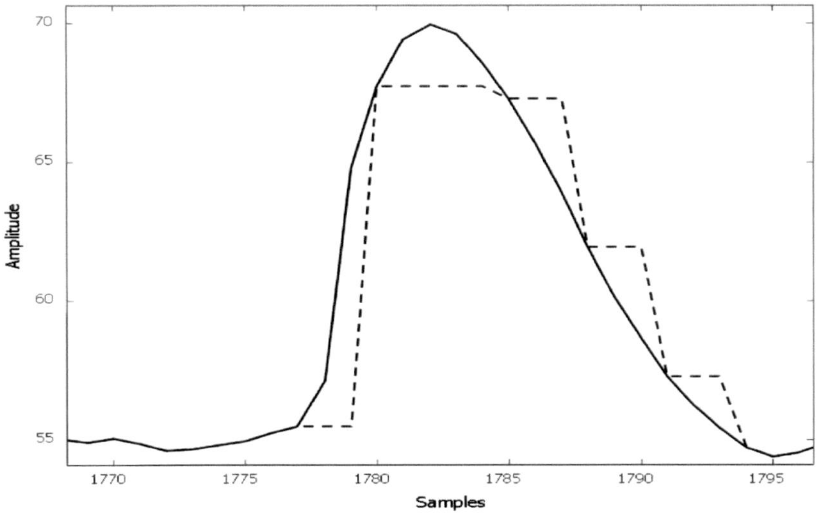

Figure 31. The wavelet filter applied to the distillation case.

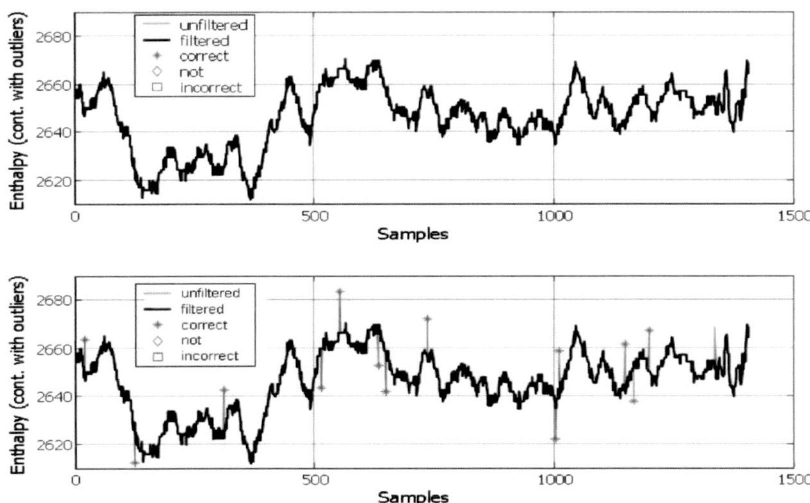

Figure 32. Boiler Data Outlier Detection: The top part, raw data with no outliers due to 2-out-of-3 set-up. The bottom part, artificially contaminated with outliers and filtered using the MAD approach plus noise level detector in the off-line mode.

same applies to the artificially contaminated data in the bottom part of Fig. 32 where all of the outliers are correctly detected and no misclassifications occur.

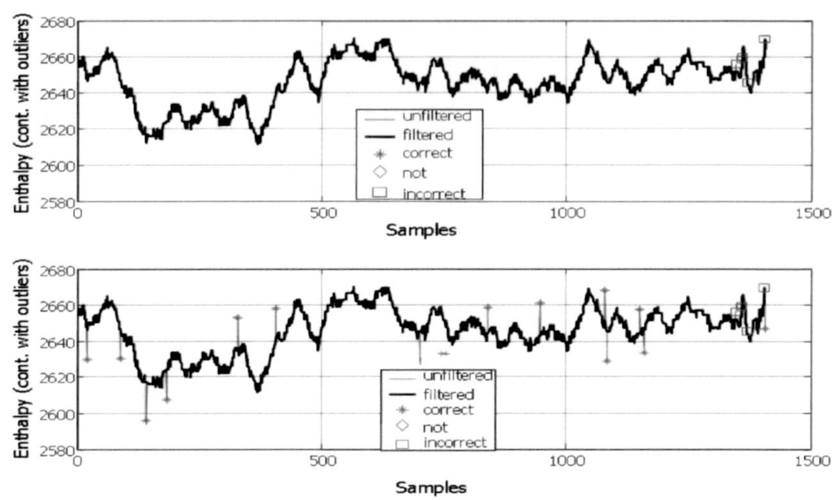

Figure 33. Boiler Data Outlier Detection: The top part, raw data with no outliers due to 2-out-of-3 set-up. The bottom part, artificially contaminated with outliers and filtered using the MAD approach plus noise level detector in on-line mode.

Fig. 33 shows the result obtained from on-line mode. The top part shows raw data from the boiler where no outliers are present. The algorithm does detect and replace some outliers, that is, the algorithm misclassifies noise as outliers at around sample number 1300. The same applies to the artificially contaminated data in the bottom part of Fig. 33 where all of the outliers are correctly detected, however, misclassifications also occur.

Distillation Column In Fig. 34, a data set from the distillation case has been contaminated with the outliers. As it can be seen the MAD algorithm performs much better than the wavelet based one. The MAD algorithm removes all the outliers and does not remove any measurements that are not outliers. On the other hand, the wavelet based approach removes some of the measurement noise and more seriously also some of the signals.

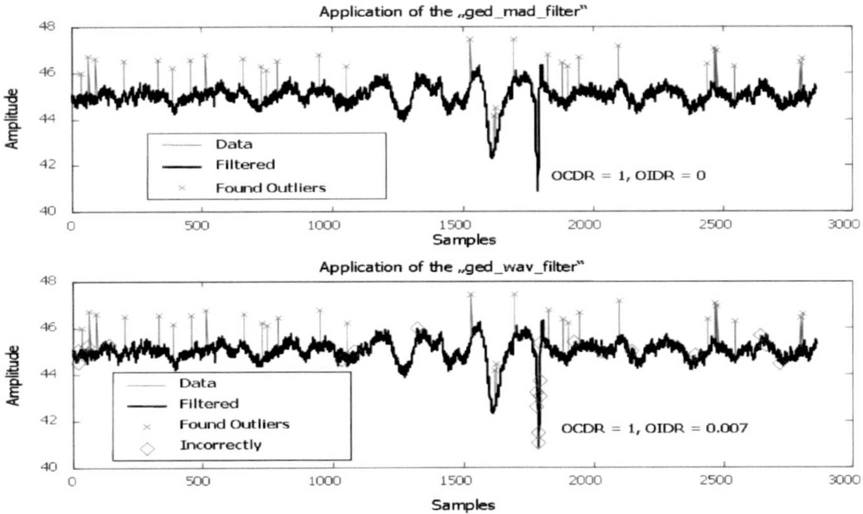

Figure 34. Applications of both algorithms on an artificially contaminated data set.

7. Summary, Conclusions and Outlook

The conclusions of this investigation are: MAD based GEDR algorithm together with the wavelet based peak noise level estimator:

- show very good performance with regard to robustness and sensitivity in the detected type 2 GEs, that is, outliers.
- show very good performance for both artificially contaminated experimental data and real contaminated experimental data.
- may be used both, for off-line and on-line applications, however, the performance disimproves in the case of an on-line application.
- The wavelet based GEDR algorithm performs well and there is still room for improvements, which should be investigated further.

References

[1] J.S. Albuquerque, L.T. Biegler (1996) Data Reconciliation and Gross-Error Detection for Dynamic Systems. *AIChE Journal* **42**: 2841–2856

[2] R.R. Coifman, M.V. Wickerhauser (1992) Entropy based algorithm for best basis selection. In *IEEE Trans. Inform. Theory,* **32**:712–718.

[3] R. Shao, F. Jia, E.B. Martin, A.J. Morris (1999) Wavelets and non linear principal component analysis for process monitoring. *Control Engineering Practice* **7**:865–879.

[4] S. Beheshti, M.A. Dahleh (2002) On denoising and signal representation. In *Proceeding of the 10th Mediterranean Control Conference on Control and Automation.*

[5] S. Beheshti, M.A. Dahleh (2003) Noise variance and signal denoising. In *Proceeding of IEEE International Conference on Acustic. Speech, and Signal Processing (ICASSP).*

[6] P.H. Menold, R.K. Pearson, F. Allgöwer (1999) Online outlier detection and removal. In *Proceeding of Mediterranean Control Conference,* Israel.

[7] R.K. Pearson (2001) Exploring process data. *J. Process Contr.* **11**:179–194.

[8] R.K. Pearson (2002) Outliers in Process Modelling and Identification. *IEEE Transactions on Control Systems Technology* **10**: 55–63.

[9] D.L. Donoho (1995) Denoising by soft thresholding. *IEEE Transaction On Information Theory,* **41** (3): 613–627.

[10] D.L. Donoho, I.M. Johnstone, (1994) Adapting to unknown smoothness via wavelet shrinkage. *Journal of the American Statistical Association,* **90** (432): 1220–1224.

[11] D.L. Donoho, I.M. Johnstone (1994) Ideal spatial adaptation by wavelet shrinkage. *Biometrika,* **81** (3): 425–455.

[12] D.L. Donoho, I.M. Johnstone, G. Kerkyacharian, D. Picard (1996) Density estimation by wavelet thresholding. *Annals of Statistics,* **24** (2): 508-539.

[13] R. Shao, F. Jia, E.B. Martin, A.J. Morris (1999) Wavelets and non linear principal component analysis for process monitoring. *Control Engineering Practice* **7**: 865–879.

[14] R. Isermann (1996) Modellgestützte Überwachung und Fehlerdiagnose Technischer Systeme. *Automatisierungstechnische Praxis* **5**: 9-20.

[15] R.R. Coifman, M.V. Wickerhauser (1992) Entropy based algorithm for best basis selection. *IEEE Information Theory* **32**: 712–718.

[16] P. Mercorelli, A. Frick (2006) "Noise Level Estimation Using Haar Wavelet Packet Trees for Sensor Robust Outlier Detection". In *Lecture Note in Computer Sciences.* Springer-Verlag publishing.

[17] http://www-stat.stanford.edu/ wavelab/

In: Fault Detection Theory, Methods and Systems ISBN: 978-1-61728-291-1
Editor: Léa M. Simon, pp. 225-252

Chapter 5

GENERATION OF FAULT DETECTING TESTS FROM FORMAL SPECIFICATIONS BY MODEL CHECKING

Andrea Calvagna* and Angelo Gargantini†
[1]Università di Catania, Italy
[2]Università di Bergamo, Italy

Abstract

We present a technique which generates from Abstract State Machines specifications a set of test sequences capable to uncover specific fault classes. The notion of *test goal* is introduced as a state predicate denoting the detection condition for a particular fault. Tests are generated by forcing a model checker to produce counter examples which cover the test goals. We introduce a technique for the evaluation of the fault detection capability of a test set. We report some experimental results which validate the method, compare the fault adequacy criteria with some classical structural and combinatorial coverage criteria and show an empirical cross coverage among faults.

*E-mail address: andrea.calvagna@unict.it
†E-mail address: angelo.gargantini@unibg.it

1. Introduction

Model-based testing (MBT) aims to reduce the cost of testing and to increase the reliability of safety critical systems. One benefit of a formal method is that the high-quality specification it produces can play a valuable role in software testing. For example, the specification may be used to automatically construct a suite of test sequences. These test sequences can then be used to automatically check the implementation software for faults. This use of the generated tests cases is also kwokn as *conformace testing*. Specification-based testing is not widely adopted [45], while white box or program based testing is well known and used in practice: many tools support it and software developers and testers are familiar with it. In the wake of the success of program based testing, specification-based testing criteria that mimic the coverage criteria for programs have been proposed. They are generally called structural criteria because they analyse the structure of the specification and require the coverage of particular elements (like states, rules, conditions, and so on). Examples are the Modified Condition Decision Coverage (MCDC), one of the most powerful criteria used in practice, applied to Abstract State Machines [22] or the coverage of properties and assertions for a program given by using the Assertion Definition Language (ADL) as proposed by [13].

Since the aim of software testing is to demonstrate the existence of errors, selecting tests that can reveal faults is of paramount importance. The fault detection capability of structural criteria is not definitely assessed though. Recent works hypothesize some classes of faults and analyze the fault detection capability of most used criteria with respect to these classes of faults. The analysis can be formal [37, 39, 41, 44] and/or empirical [49]. The main result is that many coverage criteria cannot assure the detection of several fault classes. For instance, MCDC is unable to detect faults due to missing brackets in boolean expressions. Stronger coverage criteria have been introduced (as in [37]) with the aim to detect more faults, but still the relationship between coverage criteria and faults is not well established and it is infeasible to evaluate the effectiveness of a test criterion in general [29]. For example, it can be shown "that the fact that criterion C_1 subsumes criterion C_2 does not guarantee that C_1 is better at detecting faults [20]".

Other papers define testing criteria focusing on certain classes of faults, which model commonly committed mistakes. For instance, Weyuker et al. [49] introduce the meaningful impact strategies for boolean expressions to target specifically the variable negation fault that occurs when a boolean variable is erroneously substituted by its negation. In the same paper, they introduced a family of strategies for automatically generating test cases from Boolean expressions, of which the MAX-A and MAX-B strategies are the most powerful and subsume all others. Chen and Lau introduced three testing criteria [14]: the *Multiple Unique True Point* (MUTP), the *Multiple Near False Point* (MNFP), and the *Corresponding Unique Tree Point and Near False Point* par (CUTP-NFP)} strategies, which were integrated by Chen et al. [51] in the MUMCUT testing strategy and (1) guarantees to detect seven types of fault in Boolean expressions in irredundant disjunctive normal form as MAX-A and MAX-B, and (2) requires only a subset of the test suites that satisfy the previously proposed MAX-A and MAX-B strategies. Kaminski and Ammann [35] introduced a new extension of MUMCUT, called Minimal-MUMCUT. Minimal-MUMCUT takes into account the feasibility of the three component of MUMCUT and guarantees to detect the same types of faults with fewer test cases.

All these criteria specify also the algorithms (with some possible non determinism) which can be used for test generation. Within this framework, assessing the fault detection capability of a criterion with respect to other criteria is important, since one should choose one criterion and generate the tests from it in accordance with the expected faults, although experimental data show that resulting tests are generally effective for detecting faults in other classes. The introduction of a new fault class would require the definition of new criteria capable to detect it or the investigation (formal or empirical) of the capability of existing criteria to detect it.

In this Chpater we introduce an approach which specifically aims at detecting faults in an implementation given its specification. Specifications are Abstract State Machines which are explained briefly in Section 2.. We assume (as [17]) that implementations contain only relatively simple faults (*competent programmer hypothesis*) of the kinds introduced in Section 3. and that a test set which detects all simple faults will detect more complex faults (*fault coupling effect*). Our approach could appear similar to the *mutation analysis* [8], but it does not require any mutation at all. Instead, we introduce in Section 4. the detection condition for a fault and define adequacy criteria in terms of these detection conditions. In Section 5. we present a method which uses the detection conditions to generate and to evaluate fault detecting tests. This method is based on the counter example generation of the model checker SPIN [30]. Our approach makes the introduction of a new fault class, the generation of tests detecting these faults, and the evaluation of other tests easy to realize. In Section 6. we discuss experimental results, some of which were unexpected. Related work is presented in Section 7..

2. Preliminaries

2.1. Abstract State Machines

Even if the Abstract State Machines (ASM) method comes with a rigorous scientific foundation [10], the practitioner needs no special training to use the ASM method since Abstract State Machines are a simple extension of Finite State Machines [11], obtained by replacing unstructured "internal" control states by states comprising arbitrarily complex data, and can be understood correctly as pseudo-code or Virtual Machines working over abstract data structures. A complete introduction on the ASM method can be found in [10], together with a presentation of the great variety of its successful application in different fields as: definition of industrial standards for programming and modelling languages, design and re-engineering of industrial control systems, modelling e-commerce and web services, design and analysis of protocols, architectural design, language design, verification of compilation schemes and compiler back-ends, etc. ASM theory is the basis of several languages and tools including the Abstract State Machine Language by Microsoft [6] and the AsmGofer [47]. In this paper we will use the AsmetaL notation developed as a part of the ASMETA project[1] [23], which features a tool set around the ASMs. .

The *states* of an ASM are multi-sorted first-order structures, i.e. domains of objects with functions and predicates (boolean functions) defined on them, while the *transition relation* is specified by "rules" describing how functions change from one state to the next.

Basically, a transition rule has the form of *guarded update* "**if** *Condition* **then** *Updates*" where *Updates* are a set of function updates each one of the form $f(t_1, \ldots, t_n) := t$ which are simultaneously executed when *Condition* is true. f is an arbitrary n-ary function and $t_1, \ldots, t_n, t$ are first-order terms.

To fire this rule in a state s_i, $i \geq 0$, all terms $t_1, \ldots, t_n, t$ are evaluated at s_i to their values, say $v_1, \ldots, v_n, v$, then the value of $f(v_1, \ldots, v_n)$ is updated to v, which represents the value of $f(v_1, \ldots, v_n)$ in the next state s_{i+1}. Such

[1] http://asmeta.sourceforge.net/

pairs of a function name f, which is fixed by the signature, and an optional argument $(v_1, \ldots, v_n)$, which is formed by a list of dynamic parameter values v_i of whatever type, are called *locations*. They represent the abstract ASM concept of basic object containers (memory units), which abstracts from particular memory addressing and object referencing mechanisms. Location-value pairs (loc, v) are called *updates* and represent the basic units of state change.

A more general schema is the conditional rule of the form:

$$\textbf{if } \varphi \textbf{ then } R1 \textbf{ else } Q1 \textbf{ endif} \qquad (1)$$

The meaning is: if φ is true then execute *R1* in parallel, otherwise execute *Q1*. Rules *R1* and *Q1* can be conditional rules themselves, simple updates, or other kinds of rules. There is a limited but powerful set of *rule constructors* that allow to express simultaneous parallel actions (`par`) or sequential actions (`seq`). Appropriate rule constructors also allow non-determinism (existential quantification `choose`) and unrestricted synchronous parallelism (universal quantification `forall`). A *computation* of an ASM is a finite or infinite sequence $s_0, s_1, \ldots, s_n, \ldots$ of states of the machine, where s_0 is an initial state and each s_{n+1} is obtained from s_n by firing simultaneously all of the transition rules which are enabled in s_n. The (unique) *main rule* is a transition rule and represents the starting point of the computation. An ASM can have more than one *initial state*. A state s which belongs to a computation starting from an initial state s_0, is said *reachable* from s_0.

For our purposes, it is important to recall how functions are classified in an ASM model. A first distinction is between *basic* functions which can be *static* (never change during any run of the machine) or *dynamic* (may be changed by the environment or by machine *updates*), and *derived* functions, i.e. those coming with a specification or computation mechanism given in terms of other functions. Dynamic functions are further classified into: *monitored* (only read, as events provided by the environment), *controlled* (read and write (i.e. updated by transaction rules)), *shared* and *output* (only write) functions.

2.2. Test Sequence

Adapting to ASMs some definitions common in literature for state transition systems [3, 42], we define a *test sequence* or *test* as follows.

Definition 1. A **test sequence** or **test** for an ASM $\mathcal{M}$ is a finite sequence of states (i) whose first element belongs to a set of initial states, (ii) each state follows the previous one by applying the transition rules of $\mathcal{M}$.

A test sequence ends with a state, which might be not final, where the test goal is achieved. Informally, a test sequence is a partial ASM run and represents an expected system behavior.

Definition 1 assumes the use of ASM specification as test oracle, since states supply expected values of outputs. The importance of test oracles is well known, since the generation of the sole inputs (often called *test data*) does rise the problem of how to evaluate the correctness of the observed system behavior.

Other approaches [9] consider a test as a simple pair of two consecutive states (often called *test vector*). The generation of state vectors is easier than the generation of test sequences, but test vectors are less useful. A test vector does not give the user the complete scenario of system execution, and does not provide the tester with any information about the reachability of the pair of states from a valid initial state – a pair could be not reachable at all, and even if it is reachable, the input sequence necessary to lead the system to the first state of the pair is not provided –. Note that obtaining test vectors from a test sequence is straightforward.

When the tester has access to the internals of the unit under test (UUT), inputs, outputs, and internal functions provided by test sequences are used; when the tester can observe only UUT inputs and outputs, values of internal functions are ignored. For these reasons, our definition of test makes our approach as general as possible, widely applicable, and neutral with respect to the actual use of the generated test sequences.

We define a collection of test sequences as follows.

Definition 2. A **test set** or **test suite** is a finite set of test sequences.

Given a predicate P over an ASM state we say that a test sequence t covers P, if t contains a state such that P is true in that state.

2.3. Coverage Criteria for ASMs

Structural Coverage Criteria

We summarize the following coverage criteria, originally presented in [22]. They are compared in Section 4 with the new fault based coverage criteria.

Basic rule (BR) coverage requires that for every guard (decision) there exists a test which covers the case when the decision is taken (the guard is evaluated true at least in one state belonging to a test sequence) and when the decision is not taken (the guard is evaluated false).

MCDC requires the classical modified condition decision coverage in the masking form [15] to every guard in the ASM.

Complete rule (CR) coverage requires that for every rule, its guard is evaluated true at least once and at least one update in the rules is not trivial.

Update coverage (UC) requires that for every update (in every rule) there exist a test sequence in which the update is applied and is not trivial.

BR and MCDC can be classified as (model-based) *control oriented* coverage criteria [52] as they consider only the control flow of the model, while CR and UC can be classified as data flow coverage criteria, since they consider the value of variable before its assignment to a possibly new value (this kind of an update can be considered a new *definition*). Note that MCDC implies BR and UC implies CR.

Combinatorial Coverage Criteria for ASMs

Structural based testing criteria require the formal specification of the whole system. Modeling activities may become extremely expensive and time consuming, and the tester may decide to model (at least initially) only the inputs and require they are sufficiently covered by tests. On the other hand, unintended interaction between input parameters can lead to incorrect behaviors which may not be detected by traditional testing [40, 50].

To this aim, combinatorial interaction testing (CIT) techniques can be effectively applied in practice. CIT consists of employing combination strategies to select values for inputs and combine them to form test cases. The tests can then be used to check how the interaction among the inputs influences the behavior of the original system under test. The most used combinatorial testing approach is to systematically sample the set of inputs in such a way that all t-way combinations of inputs are included. This approach exhaustively explores t-strength interaction between input parameters, generally in the smallest possible test executions. This approach has been implemented for ASM in [12].

3. Fault Classes

While test coverage criteria like the CR and UC, presented in Section 2.3., aim to detect faults in updates, in this paper we focus only on faults which may occur in the guards of the ASM specification under test. Note that a *fault* in a implementation is a cause that results in a failure [34], which is an erroneous evaluation of a guard in the implementation in our approach. We consider only faults originated by typical programmer mistakes like use of incorrect control predicates, missing conditions, and the incorrect use or order of boolean and relational operators in rule guards. These types of mistakes result in faults that regard the conditions and their operators, where with *condition* we intend atomic boolean expressions which cannot be further decomposed in simpler boolean expressions. A condition can be a boolean variable, like `overridden`, or a relational expression like `pressure > TooLow`. We exclude faults inside conditions except the incorrect use of relational operators (for instance the use of > instead of <). We follow the notation proposed in [41]: a *literal* [2] is an occurrence of a condition inside a guard (note that a condition or a boolean variable may occur several times in the same guard). While many papers [39, 48, 41] assume that the boolean expressions are given in disjunctive normal form (DNF), we remove such restriction (as in [44]). We study the following fault classes:

- Operand faults (i.e. regarding the literals or sub expressions):

LNF *Literal negation fault.* An occurrence of a condition (i.e. one literals) is replaced by its negation.

ENF *Expression negation fault.* It consists of replacing a sub expression (but not a condition or literal) with its negation.

MLF *Missing literal fault.* It causes the absence of one literal or condition in the formula. If the same condition occurs several times in the formula, we introduce several faults and not just one.

ST0/1 *Stuck at 0/at 1 fault.* This is a classical hardware fault, which consists in replacing an input with 0 or with 1. In our case, it causes a replacement of a literal by *false* (ST0) or *true* (ST1). ST0U1 denotes the union of ST0 and ST1, i.e. the replacement of a literal by false and true.

- Boolean Operator faults:

[2]A literal is sometimes called *clause* as in [44], a condition is often called *variable* especially in papers dealing with boolean specifications [41, 39, 48].

Table 1. Examples of faults

Fault Class	Faults for $a \wedge b \wedge (a \vee b)$
LNF	$\neg a \wedge b \wedge (a \vee b)$, $a \wedge \neg b \wedge (a \vee b)$, $a \wedge b \wedge (\neg a \vee b)$, $a \wedge b \wedge (a \vee \neg b)$
ENF	$\neg(a \wedge b \wedge (a \vee b))$, $\neg(a \wedge b) \wedge (a \vee b)$, $a \wedge b \wedge \neg(a \vee b)$
MLF	$b \wedge (a \vee b)$, $a \wedge (a \vee b)$, $a \wedge b \wedge b$, $a \wedge b \wedge a$.
ST0	$false \wedge b \wedge (a \vee b)$, $a \wedge false \wedge (a \vee b)$, $a \wedge b \wedge (false \vee b)$, $a \wedge b \wedge (a \vee false)$
ASF	$(a \wedge b \wedge a) \vee b$
ORF	$a \wedge b \vee (a \vee b)$, $(a \vee b) \wedge (a \vee b)$, $a \wedge b \wedge (a \wedge b)$.

ASF *Associative Shift fault.* This fault is due to the misunderstanding about operator evaluation priorities and missing brackets.

ORF *Operator Reference fault.* $'\wedge'$ is replaced by $'\vee'$ and vice-versa[3].

Furthermore we add the following fault class, which introduces faults in relational expressions with pattern $E\ op\ F$, where E and F are either arithmetic expressions or expressions of enumerative type and *op* is one of $<$, $\leq$, $=$, $>$, $\geq$, and $\neq$.

ROF *Relational Operator Fault.* Replace a relational operator by any other relational operator (note that replacing an operator with its opposite is equal to LNF). If the expression is an enumeration, then replace only $=$ with $\neq$ and vice-versa (we allow only "equals" and "differs" comparison between two enumerative values). For example, from $x \leq c$ one would obtain the following faulty expressions: $x > c$, $x \geq c$, $x = c$, $x \neq c$, and $x < c$.

We have chosen LNF, ENF, MLF, ASF, and ORF because they are the most studied faults in the literature [39, 48, 41]. ST0/1 faults are commonly considered a realistic model of manufacturing faults in hardware circuit testing. ROF is introduced and studied in [44] (called Relational Operator Reference Fault)

[3] We assume that logical binary operators are left associative, hence $a \wedge b \wedge (a \vee b)$ must be read as $(a \wedge b) \wedge (a \vee b)$

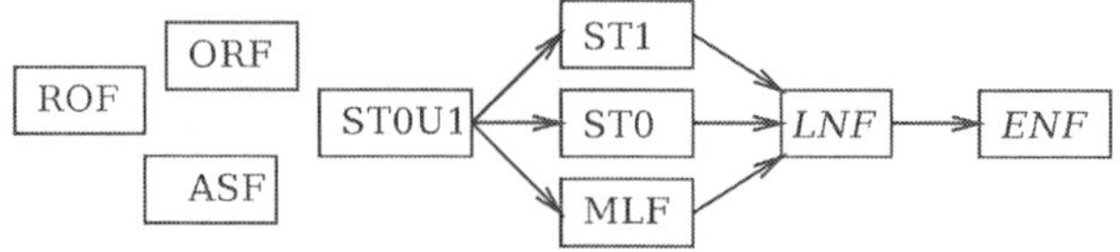

Figure 1. Boolean fault hierarchy.

and models a typical software fault. Faults of omission (modeled by the MLF class) are known to be very common, constituting approximately half of the bug reports posted on Usenet [39]. In a recent realistic case, Dupuy and Leveson examined an attitude control software for the HETE-2 (High Energy Transient Explorer) scientific satellite and uncovered an important operator reference fault (ORF) which replaced an AND operator by an OR [18].

The hierarchy among fault classes for boolean expressions have been intensively studied. For instance, empirical work [49] showed that tests generated to detect variable negation fault (our LNF) always detected expression negation faults. Kuhn proposed a rigorous approach to formally prove the existence of a hierarchy among faults in boolean specifications given in normal form [39]. The initial hierarchy proposed by Kuhn was first enriched by [48] and then by [41]. Okun et alt. [44] developed a novel analytic technique to find the hierarchy among faults of arbitrary boolean expressions, not just those in disjunctive normal form. According to the results presented in the literature, the hierarchy among the fault classes used in this paper is presented in Figure 1, where $C_1 \rightarrow C_2$ means that every test suite able to detect C_1 detects C_2 as well. In this case, we say that C_1 is *stronger* than C_2.

Note that a pair of fault classes C_1 and C_2 is proved to be independent in the hierarchy when there exists a test suite which guarantees to detect faults of C_1 but not those of C_2 and vice-versa. In our case, ORF, ROF, and ASF are independent of each other and with all the other fault classes, and MLF, ST0, and ST1 are independent of each other. This fact has practical consequences: since a test set T_1 which detects a fault C_1 does not guarantee to detect C_2 and vice-versa, one should generate a test suite for C_1 and a test suite for C_2. Therefore, one should generate a test suite for every independent fault class. However, T_1 may detect C_2 as well for the particular specification under test and the generation for C_2 may be skipped. To assess the actual fault detection capability of a test suite, we introduce in Section 5. a method to evaluate tests with respect to possible faults in the specification under test and regardless of the way such tests have been generated.

4. Discovering Faults

The erroneous implementation of a boolean expression φ as φ' can be discovered only when the expression $\varphi \oplus \varphi'$, called *detection condition,* evaluates to true, where $\oplus$ denotes the logical *exclusive or* operator. Indeed, $\varphi \oplus \varphi'$ is true only if φ' evaluates to a different value than the correct predicate φ. The detection condition is also called *boolean difference* or *derivative* [1].

Consider a simple rule R of an ASM specification $\mathcal{M}$:

R = **if** φ **then** *updates*

Let $\mathcal{M}'$ be the faulty implementation of $\mathcal{M}$. Assume that the guard φ of R in $\mathcal{M}$ is erroneously implemented as φ' in $\mathcal{M}'$ due to the fault F, and that rule *updates* are not all trivial. F can be detected during testing only if there exists a test sequence t containing a state s in which $\varphi \oplus \varphi'$ is evaluated to true, i.e. φ and φ' have different values in s for $\mathcal{M}$ and $\mathcal{M}'$. In this case, when we apply t, the rule R fires in $\mathcal{M}$ and performs its updates but it does not fire in $\mathcal{M}'$ or vice-versa. Note that in this paper we assume that in oder to find a fault, it suffices that the test drives the faulty system to a different state with respect to that required by the specification. This assumption is similar to that done in weak mutation testing. A distinction exists between *weak* and *strong* mutation testing. In strong mutation testing, a fault is found if and only if the final output of the faulty and original program differ. In weak mutation testing [33], this is relaxed so that a test is considered sufficient if the program state of the mutant and original program differ after execution of the mutated statement. If a mutant is weakly killed, the mutant and original program can still have identical outputs. In practice, weak mutation testing is almost as effective as strong mutation testing, with major computational savings [43]. That is, most mutants that are weakly killed are strongly killed.

Let φ be the guard of R and φ' its implementation due to the fault F.

Definition 3. Detection condition. The predicate $\varphi \oplus \varphi'$ is the *detection condition* of F and it is called *test predicate* or test goal.

Example 4. If the guard $x \leq c$ in the specification is implemented as $x < c$, then the test goal is $x \leq c \oplus x < c$ which is equivalent to $x = c$. Only a test sequence containing a state s in which $x = c$ can uncover the fault.

Let φ be a guard and C be a fault class. We denote with $F_C(\varphi)$ the set of all the possible faulty implementations of φ according to the fault class C (as explained in Section 3.). The test predicates to discover the fault C in φ are the expressions $\varphi \oplus \varphi'$ with φ' in $F_C(\varphi)$. For example, if the guard is $a \wedge b$ and C is the MLF, then $F_{MLF}(a \wedge b) = \{a, b\}$ and the test predicates are the following two expressions: $(a \wedge b) \oplus a$ (which is $a \wedge \neg b$) and $(a \wedge b) \oplus b$ (which is $\neg a \wedge b$).

In case of a nested rule of kind (2), test predicates must include the guards of outer rules. Let φ be the guard of an inner rule R and $g_1, \ldots, g_n$ be the guards of the outer rules or their negation (in case of else) such that if $g_1 \wedge \ldots \wedge g_n$ holds, then R is executed (and its updates fired if φ is true). We call $g_1, \ldots, g_n$ *outer guards* of R. The test predicates to discover the fault C in φ are the expressions $g_1 \wedge \ldots \wedge g_n \wedge (\varphi \oplus \varphi')$ with φ' in $F_C(\varphi)$.

Definition 5. Test Predicates. Let R be a rule in an ASM $\mathcal{M}$, φ be its guard, $g_1, \ldots, g_n$ be the outer guards of R, and C be a fault class. The set $\Gamma_C(R)$ of test predicates is given by the expressions $g_1 \wedge \ldots \wedge g_n \wedge (\varphi \oplus \varphi')$ with φ' in $F_C(\varphi)$.

A test suite is adequate to test the guard φ of a rule R with respect to a fault class C if it covers every test predicate generated for R and C:

Definition 6. Fault Detecting Adequacy Criteria. A test suite $\mathcal{T}$ is adequate with respect to the fault class C and the ASM $\mathcal{M}$, if for every rule R of $\mathcal{M}$ and for every test predicate $\mathtt{tp}$ in $\Gamma_C(R)$ there exists a state s in a test sequence of $\mathcal{T}$ such that the test predicate $\mathtt{tp}$ evaluates to true in s.

5. Generation of Tests

The generation of tests can be performed by means of model checking algorithms by exploiting the counterexample generation capability of these algorithms. Model checkers have been successfully applied to formal verification of properties, normally given in temporal logic, for systems modeled by means of automata. Model checking allows generation of finite paths for logical predicates, either showing satisfaction or violation (*counterexample*). To do so, the test predicates have to be formalized as temporal logic formulas. In most cases counterexample generation is exploited because it is supported by all model checking techniques. Model checkers automatically perform the proof of a desired property p by analyzing every possible system behavior, checking that p

is true, and producing a counter example in case the property p does not hold in the model. The counter example is a possible system behavior that shows a case where the property p is falsified. In order to force the model checker to generate a counterexample the test predicates are formalized in a negated fashion, such that the counterexample satisfies the original test predicate. Such properties are known as *trap properties*. In this chapter, we exploit the capability of the model checker Spin [30] to produce counter examples. For a survey on the use of model checkers for test generation see [21].

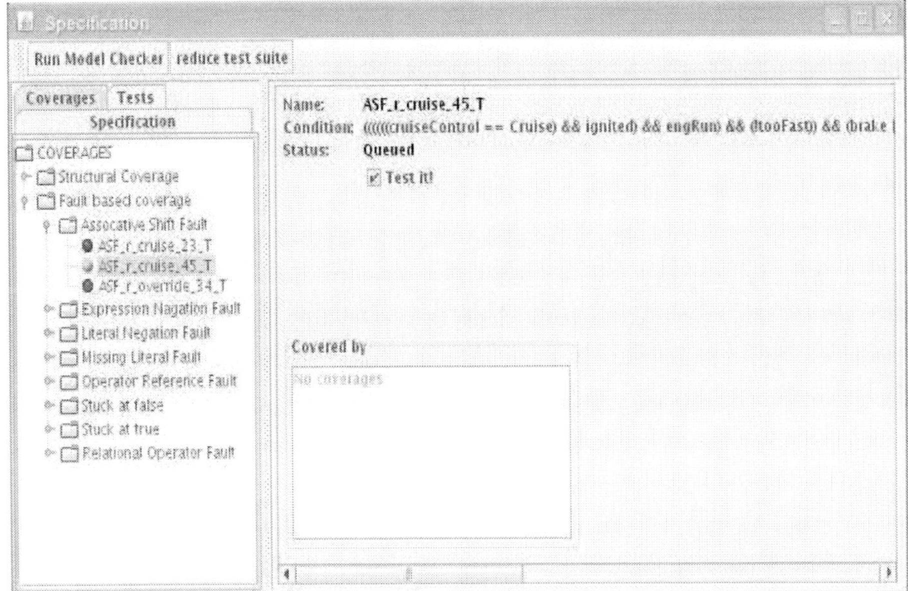

Figure 2. An ATGT screen-shot.

The method presented in this section has been implemented in a prototype tool ATGT[4] - a screen-shot is reported in Fig. 2 - and consists in the following steps as illustrated in Figure 3.

- First, denoted by ①, a *Test Predicate Generator* computes the test predicate set $\Gamma_C = \{tp_i\}$ for the desired fault classes C introduced in Section 3. and for all the rule guards in the ASM specification under test given in the syntax of the AsmGofer [47].

[4]available at http://cs.unibg.it/gargantini/projects/atgt/

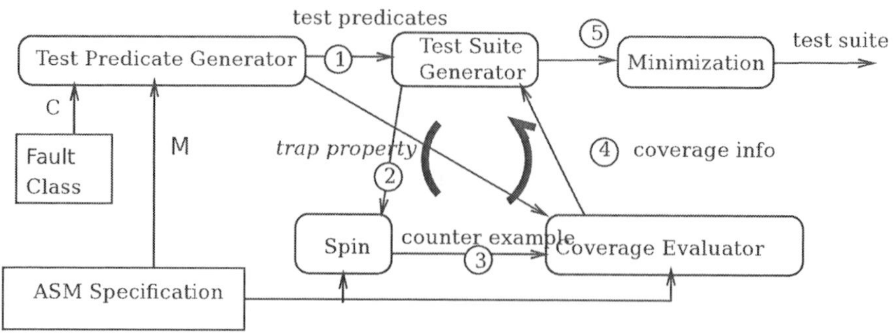

Figure 3. Steps in the proposed generation method.

- Second ②, the *Test Suite Generator* selects a test predicate tp_i, either randomly or according to the user request, and computes the *trap property* stating that tp_i is never true, i.e. $\mathsf{p} = \mathbf{never}(tp_i)$ which is translated in PROMELA, the language of Spin, as the statement `assert(!tp_i)`. The trap property p is not a desired property of the system; on the contrary, we look for a system behavior which falsifies p, i.e. where tp_i becomes true. This method has been introduced in [24, 25].

- Third ③, the model checker is used to find the test sequence, by encoding the ASM specification in PROMELA, following the algorithms described in [25] and trying to prove the trap property p. If the model checker finds that p is false, i.e. a state where the tp_i is true, it stops and prints as counter example the state sequence leading to that state (plus the updates generated by the last update). This sequence represents the test that covers tp_i.

- Fourth ④, the *Coverage Evaluator* reads the counter example to produce the actual test sequence and to evaluate its coverage as explained in the following section against all the test predicates generated in the first step and provides the coverage information back to the Test Suite Generator.

- The process is iterated starting from the second step for each test predicate that has not been already covered. Once all the test predicates have been covered, the obtained test suite is reduced by minimization ⑤, which removes unnecessary tests from the final test suite.

- In the end, a complete test suite is generated, except for the cases where the model checker fails to find a counter example as explained in Section 5.1.

Coverage Evaluation and Monitoring

A test sequence that is generated to cover a particular test predicate likely covers also many other test predicates, i.e. it contains a state where other test predicates are true and it can, therefore, discover other faults as well. Finding which predicates are covered (also called *monitoring*) can reduce the time and the resources to obtain a complete test suite - because we can decide to skip the generation of tests for test predicates already covered - and it can reduce the size of the test suite - because we could decide to discard a test if the test predicates that it covers are covered also by other tests -. Note that the evaluation of coverage can be used to evaluate any test sequence, regardless the way it has been generated. As we show in Section 6., we use this technique to get valuable insights over the fault detection capability of the structural coverage criteria presented in Section 2.3.

Minimization

A test suite is minimal [27] with regard to an objective if removing any test case from the test suite will lead to the objective no longer being satisfied. The problem of finding the optimal (minimal) subset is NP-hard, which can be shown by a reduction to the minimum set covering problem. ATGT uses a simple greedy heuristic to the minimum set covering problem for test suite minimization: The heuristic selects the test case that satisfies the most test predicates and remove all test predicates satisfied by that test case. This is repeated until all test predicates are satisfied.

Monitoring and minimization can behave very differently: Minimization requires existing, full test suites while monitoring checks test predicates on the fly during test case generation. On the other hand, monitoring does not guarantee minimal test suites.

Note that the post reduction may reduce the fault detection capability of the test suite, but not with respect the fault classes it initially covered since the set of test predicates covered remains the same.

5.1. Undetectable Faults

When the model checker terminates, one of the following three situations occurs. The best case occurs when the model checker stops finding that the trap property is false, and, therefore, the counter example that covers the test predicate is generated.

The second case happens when the model checker checks every possible behavior without finding any state where the trap property is false, and, therefore, it actually proves the trap property **never**(tp_i). A test predicate, in our case, has always the pattern $A \wedge \varphi \oplus \varphi'$ (where A is a conjunction of outer guards), and **never**$(A \wedge \varphi \oplus \varphi')$ is equivalent to **always**$(A \rightarrow \neg(\varphi \oplus \varphi'))$, i.e. **always**$(A \rightarrow (\varphi \leftrightarrow \varphi'))$. Therefore, SPIN proves that when A holds, φ is always equivalent to its mutation φ' and that the fault does not introduce an actual change in the behavior of the system. In this case we say that such a fault is undetectable and we can safely ignore tp_i and simply warn the user that its model is insensitive in that rule guard to that fault.

In the third case, the model checker terminates because it finishes the maximum time or memory allocated for the search (set by the user or decided by model checker itself) but without completing the state space search and without finding a violation of the trap property, and, therefore, without producing any counter example (generally because of the state explosion problem). In this case, we do not know if either the trap property is true (i.e., the fault cannot be discovered) but too difficult to prove, or it is false but a counter example is too hard to find (i.e. the fault could be discovered if an appropriate test sequence could be found). When this case happens, our method simply warns the tester that the test predicate has not been covered, but it might be feasible.

Model Checking Limits. Model checking applies only to finite models. Therefore, our method works for ASM specifications having variables and functions with finite domains. The problem of abstracting models with finite domains from models with infinite domains such that some behaviors are preserved, is under investigation. Moreover, since model checkers perform an exhaustive state space (possibly symbolic) exploration, they fail when the state space becomes too big and intractable. This problem is known as *state explosion problem* and represents the major limitation in using model checkers. Note, however, that we use the model checker not as a prover of properties we expect to be true, but to find counter examples for trap properties we expect to be false.

Therefore, our method does generally require a limited search in the state space and not an exhaustive state exploration. However, undetectable faults require a complete state search.

Model Checking Benefits. Besides its limits, model checking offers several benefits. For instance, SPIN adopts sophisticated techniques to compute and explore the state space, and to find property violations. It represents a state and the state space in a very efficient way using state enumeration, hashing techniques, and state compression methods. Moreover, SPIN explores the state space using practical heuristics and other techniques like partial order reduction methods and on-the-fly state exploration based on a nested depth first search. For these reasons, we have preferred existing model checkers instead of developing our own tools and algorithms for state space exploration. Moreover, the complete automaticity of model checkers allows to compute test sequences from ASM specifications without any human interaction.

6. Experiments

We report the result of applying our method to two case studies, the Cruise Control (CC) specification [41, 5] and a simple model for a Safety Injection System (SIS) of a nuclear plant [16, 22, 24]. The CC has one monitored (i.e. modified only by the environment) enumerative variable, 4 monitored boolean variables and one controlled (i.e. modified only by the system) variable. It has 9 rules with rather complex boolean expressions as guards, which admit numerous boolean operator faults. The SIS includes three monitored variables (one integer in the interval [0,2000] and two switches), two internal variables (a boolean and an enumerative) and an output (boolean). It has 7 transition rules with guards which contain several relational operators and hence admit numerous ROFs. The number of test predicates is shown in Table 2. Note that 1 test predicate for SIS and 20 test predicates in the CC for the ROF were proved unfeasible by the model checker, which completed the search without finding any violation of the trap property, therefore actually proving that the faults are undetectable as explained in the second case of Section 5.1..

Table 2. Test Predicates and Tests for SIS and CC

#tp	ASF	ENF	LNF	MLF	ORF	ST0	ST1	ROF (unfeasible)
for SIS	1	9	16	16	9	24	24	15 (1)
for CC	0	22	29	28	22	49	49	29 (20)

6.1. Generation of Tests

We have applied three strategies for test generation. In strategy 1 and 2 we use the breath first search (BFS) algorithm of Spin, which normally requires more time and memory than the default nested depth first search (nDFS) algorithm, but it guarantees that the shortest counter example is found. In strategy 3 we use the nDFS which is faster but finds long counter examples. Furthermore, in the first strategy we start from weaker fault classes and then we increase the fault detection capability of the tests by choosing stronger faults, while in the second and third strategy we start from strong coverage classes. Results are shown in Table 3, in which we report the number of runs, the total time required[5], the number of tests (some tests are discarded because they cover only test predicates covered by other test sequences in the test suite), and the total number of states in the test sequences.

Table 3. Runs for test generation

strategy	#runs	time (sec)	#test	#states
1 - BFS, weak to strong	59	116	22	635
2 - BFS, strong to weak	59	102	22	637
3 - DFS, strong to weak	42	258	8	11760

Although several papers [39, 41, 36] suggest that hierarchical information about fault classes can be useful during test generation and that starting the test generation from the strongest coverage would require less time and fewer test cases than starting from the weakest coverage, we found no evidence of this fact. Indeed, strategy 2 (strong to weak) performed as well as strategy 1 (weak to strong). This result can be explained by considering that our method is iter-

[5]We have used a PC with an AMD Athlon 3400+ and 1 GB of RAM

ative (it produces a test sequence at a time) and that we perform test evaluation at the end of every cycle. If the criterion S is stronger than the criterion W, any test set T_S adequate according to S includes any set of test T_W adequate according to W. The test generation starting from S produces a test suite T_S, whose evaluation stops the test generation because T_S covers W as well. The test generation starting from W initially produces T_W which still requires the generation of $T_S - T_W$ and not of the complete T_S. In both cases the number of test cases is the same (except for some non determinism in the generation and in the optimization of the test suites). However, our examples are too small to draw the definitive conclusion that hierarchical information about fault classes are useless during test generation.

Another unexpected result was that the test generation with the DFS algorithm performed worse than the others, although the DFS proved to be more efficient per visited state than the others: it explored around 11760 states (18 times more then the others) but it took only about twice as much time. By analyzing the runs, we found that the sole model checker execution (step ③ in Fig. 3) actually took less time than the same step in other strategies, but the other steps which analyze the results to evaluate the coverage took much more time, since the DFS produces very long counter examples. We believe that strategy 3 may perform better than the others for complex specifications, since in complex cases the model checker execution is the most critical step in the proposed test generation method. Moreover, strategy 3 is useful when one prefers very few test cases (for example if resetting the system is expensive) and because long test sequences may discover more faults (like extra states) [42, 28].

6.2. Comparison with Structural and Combinatorial Coverage Criteria

We have compared our new fault based adequacy criteria and the structural criteria presented in Section 2.3.. Tests for structural criteria are generated following the technique introduced in [22, 25]. Table 4 reports the structural coverage of tests generated to cover faults. Only the ST0 (and ST0U1) has covered all the structural parts in our specification. MLF test set has covered all except the Update Coverage. Table 4 reports also the combinatorial coverage of fault based test suites. No fault based test suite was able to achive all the combinatorial coverages, with MLF and ST0U1 scoring better than the others.

Table 4. Structural coverage of fault criteria (in %)

	Structural Coverage					Combinatorial Coverage				
	BR	**CR**	**UC**	**MCDC**	$\sum$	**pair**	**3wise**	**4wise**	**5wise**	$\sum$
ASF	21.4	0	10	21.4	15.4	15.6	8.3	0	0	12.5
ENF	71.4	0	50.7	74.6	62	68.8	52.2	60.7	50	56.1
LNF	92.9	*100*	92.9	94.4	93.5	90.6	78.4	67.9	54.2	78.3
MLF	*100*	*100*	95	*100*	98.1	*100*	*100*	*100*	97.9	99.8
ORF	78.6	0	57.9	80.2	68.5	68.8	52.2	60.7	50	56.1
ST0	*100*	*100*	*100*	*100*	*100*	93.8	87.5	*100*	97.9	90.9
ST1	92.9	*100*	87.9	94.4	91.6	96.9	90.9	67.9	54.2	87.2
ST0U1	*100*	*100*	*100*	*100*	*100*	99	*100*	*100*	97.9	99.7
ROF	85.7	*100*	85.7	88.9	87	84.4	73.1	60.7	50	73.9

6.3. Cross Coverage Among Fault Classes

We have also analyzed the *cross coverage* among the fault based criteria and results are reported in Table 5, which must be read as follows. The tests generated for a fault class in a row (cross) covers also the shown percentage of the test predicates for other fault classes displayed in the columns. Besides the confirmation of the theory (continuous arrows in the figure of Table 5), we have found some empirical relationships among fault classes (dotted arrows in the figure). For instance, ROF covered all the ENFs, ORF covered all the ENFs, and LNF covered all ORFs. MLF seems stronger than ST1 and ST0 individually, and ROF seems stronger than ENF and ORF. Although this empirical extended hierarchy may not hold in general, we believe that for most boolean expressions these relationships are likely to be true. This information may be useful in practice if one has a test suite that targets a specific fault and want to approximately judge the test suite's fault detection capability.

7. Related Work

Many papers tackle the problem of tests generation or selection. The subject of using model checkers for test generation starting from models has been studied for many years. For a (not so recent) survey see [2]. In [19] the authors used Spin to generate test sequences for a protocol augmented by a test pred-

Table 5. Cross Coverage (%)

	ASF	ENF	LNF	MLF	ORF	ST0	ST1	ROF
ASF	-	44	31	16	33	19	27	32
ENF	100	-	77	42	84	62	65	87
LNF	100	100	-	72	100	86	87	98
MLF	100	100	100	-	100	100	100	100
ORF	100	100	85	44	-	71	70	87
ST0	100	100	100	83	100	-	86	100
ST1	100	100	98	86	100	86	-	98
ST0U1	100	100	100	100	100	100	100	100
ROF	100	100	86	67	95	77	79	-

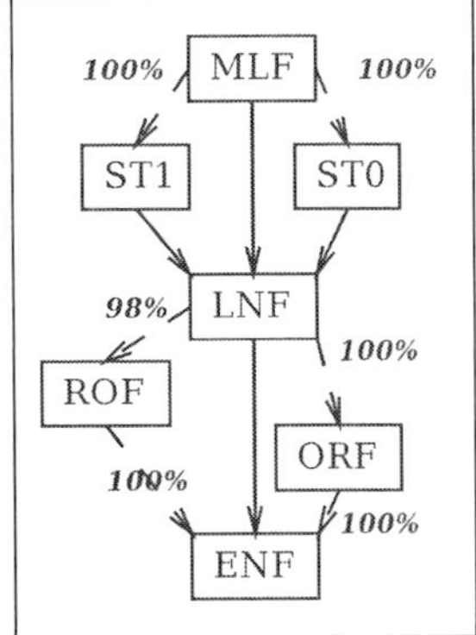

icate, called test purpose, written by the designer by hand. Classical control oriented tests generation is presented for SCR in [24] and for ASMs in [22]. Several recent papers apply the same concepts to UML state diagrams [38], to StateCharts [32], and to Stateflow [26] specifications. [46] presents state coverage, decision coverage and MCDC (not masked) for specifications written in RSML^{-e}. They all share the same approach. They introduce some control oriented coverage, derive the test predicate from decision points in the model and then use the model checker to obtain the test sequences.

A first attempt to introduce data flow oriented coverage criteria can be found in [22] where the rule update criterion (presented in Section 2.3.) covers the real update of a variable. A novel approach is presented in [31], which shows how the classical data flow coverage criteria can be translated in terms of the Computation Tree Logic (CTL).

The combined use of model checking and mutation testing is presented in [3, 8]. Their approach, that we could classify as fault oriented [52], is very similar to ours, but the technique is completely different. Differently from us, they do not use test predicates derived from the specifications by using the boolean difference. Instead, they directly apply mutation techniques to models. The original specification, written for the model checker SMV, is initially augmented by many temporal logic properties (*constraints*) that represent the correct behavior. In the extraction of these properties (also called *expounding*) there are several "subtle issues that require attention [4]" and may reduce the fault detection capability of the tests. Afterwards, the specification or the constraints are repeatedly modified applying mutation operators (more general than our fault classes), that introduce faults in the models or in the constraints. Counter examples are automatically generated by SMV either (*approach 1*) trying to prove the original properties in faulty models to obtain wrong behaviors that implementations must not exhibit or (*approach 2*) proving mutated properties for the correct model to obtain tests sequences that discover particular faults (or *kill* mutants).

We can compare their approach 2 with ours as follows. In the extraction of constraints, they build a set of safety properties which are always true in the original model. Given a safety property $\mathbf{always}(P)$, they look for a counter example by trying to prove $\mathbf{always}(P')$ where P' is a possible mutation of P. If a counter example is found, they have found a state where the mutated property is false, i.e. $\neg P'$. They actually find a state where P is true (safety property) while P' is false, i.e. $P \wedge \neg P'$, which is a particular case of $P \oplus P'$, the boolean difference of P. Our approach does not require the extraction of safety properties, since test conditions are defined as boolean differences over guards, which are not always true.

Ammann et alt. tackle also the problem of evaluation of test sequences against specification-based coverage criteria [4]. They show how the model checker SMV can be used to evaluate a test sequence with respect to the capability to discover (or *kill*) mutations of the original specification. The test sequence (regardless the way it has been generated) is transformed in a SMV model to run together with the mutated specification. This requires a run for every test and every mutation, rising the problem how to reduce (*winnow*) the number of mutations really necessary to evaluate the coverage of a test. Tests which kill a subset of mutations of other tests, can be discarded. In our approach, we can evaluate the capability of a test sequence to detect all faults in one run by using test conditions. Furthermore we are able not only to discard duplicated test cases, but also to avoid the generation of tests for test predicates already covered.

Model checkers can be used to generate tests in program based testing too: the model checker BLAST is used in [7] to generate test suites and to detect dead code in C programs.

8. Conclusion

We have presented a technique for generation of test sequences from Abstract State Machines specifications and these tests are capable to uncover specific fault classes. The dectection condition for a fault is formalized by a predicate called test goal or test predicate. Tests are generated by forcing a model checker to produce counter examples which cover the test goals. Experimental results show that the method is viable and the model checker can efficiently generate the test suites.

References

[1] S. B. Akers. On a theory of boolean functions. *Journal Society Industrial Applied Mathematics*, **7**(4):487–498, December 1959.

[2] Paul Ammann, Paul E. Black, and Wei Ding. Model checkers in software testing. Technical Report NIST-IR 6777, National Institute of Standards and Technology, 2002.

[3] Paul Ammann, Paul E. Black, and William Majurski. Using model checking to generate tests from specifications. In *2nd IEEE International Conference on Formal Engineering Methods (ICFEM'98)*, page 46, Brisbane, Australia, December 1998.

[4] Paul E. Ammann and Paul E. Black. A specification-based coverage metric to evaluate test sets. *International Journal of Reliability, Quality and Safety Engineering*, **8**(4):275–300, December 2001.

[5] Joanne M. Atlee and Michael A. Buckley. A logic-model semantics for SCR software requirements. In *ISSTA '96: Proceedings of the 1996 ACM SIGSOFT international symposium on Software testing and analysis*, pages 280–292, New York, NY, USA, 1996. ACM Press.

[6] Michael Barnett and Wolfram Schulte. The ABCs of specification: AsmL, behavior, and components. *Informatica*, **25**(4):517–526, 2001.

[7] Dirk Beyer, Adam J. Chlipala, Thomas Henzinger, Ranjit Jhala, and Rupak Majumdar. Generating tests from counterexamples. In *Proc. International*

Conference on Software Engineering (ICSE), Edinburgh, pages 326–335. IEEE CS Press, May 2004.

[8] Paul E. Black, Vadim Okun, and Yaacov Yesha. Mutation of model checker specifications for test generation and evaluation. In W. Eric Wong, editor, *Mutation Testing for the New Century, proc. of Mutation 2000*, pages 14–20. Kluwer Academic Publishers, October 2000.

[9] Mark R. Blackburn and Robert D Busser. T-VEC: A tool for developing critical systems. In *Compass'96: Eleventh Annual Conference on Computer Assurance*, Gaithersburg, Maryland, 1996. National Institute of Standards and Technology.

[10] E. Börger and R. Stärk. *Abstract State Machines: A Method for High-Level System Design and Analysis*. Springer-Verlag, 2003.

[11] Egon Börger. The ASM method for system design and analysis. A tutorial introduction. In Bernhard Gramlich, editor, *Frontiers of Combining Systems, 5th International Workshop, FroCoS 2005, Vienna, Austria, September 19-21, 2005, Proceedings*, volume 3717 of *Lecture Notes in Computer Science*, pages 264–283. Springer, 2005.

[12] Andrea Calvagna and Angelo Gargantini. A logic-based approach to combinatorial testing with constraints. In Bernhard Beckert and Reiner Hähnle, editors, *Tests and Proofs, Second International Conference, TAP 2008, Prato, Italy, April 9-11, 2008. Proceedings*, volume 4966 of *Lecture Notes in Computer Science*, pages 66–83. Springer, 2008.

[13] Juei Chang and Debra J. Richardson. Structural specification-based testing: Automated support and experimental evaluation. In Oscar Nierstrasz and Michel Lemoine, editors, *Software Engineering – ESEC/FSE '99*, volume 1687 of *Lecture Notes in Computer Science*, pages 285–302. Springer-Verlag, November 1999.

[14] Tsong Yueh Chen and Man Fai Lau. Test case selection strategies based on boolean specifications. *Softw. Test., Verif. Reliab.*, **11**(3):165–180, 2001.

[15] John Chilenski and L. A. Richey. Definition for a masking form of modified condition decision coverage (mcdc). Technical report, Boeing, Seattle WA, 1997.

[16] P.-J. Courtois and David L. Parnas. Documentation for safety critical software. In *Proc. 15th Int'l Conf. on Softw. Eng. (ICSE '93)*, pages 315–323, Baltimore, MD, 1993.

[17] R. A. DeMillo, D. S. Guindi, K. N. King, W. M. McCracken, and A. J. Offutt. An extended overview of the Mothra software testing environment. In *Proceedings of the Second Workshop on Testing, Analysis, and Verification*, pages 142–151. IEEE Computer Society Press, 1988.

[18] A. Dupuy and N. Leveson. An empirical evaluation of the mc/dc coverage criterion on the hete-2 satellite software. In *The 19th Digital Avionics Systems Conferences, 2000. Proceedings. DASC*, 2000.

[19] Andre Engels, Loe Feijs, and Sjouke Mauw. Test generation for intelligent networks using model checking. In E. Brinksma, editor, *Proceedings of the Third International Workshop on Tools and Algorithms for the Construction and Analysis of Systems, TACAS'97*, number 1217 in Lecture Notes in Computer Science, pages 384–398. springer, 1997.

[20] Phyllis G. Frankl and Elaine J. Weyuker. A formal analysis of the fault-detecting ability of testing methods. *IEEE Transactions on Software Engineering*, **19**(3):202–213, March 1993.

[21] Gordon Fraser, Franz Wotawa, and Paul E. Ammann. Testing with model checkers: a survey. *Software Testing, Verification and Reliability*, **19**(3):215–261, 2009.

[22] A. Gargantini and E. Riccobene. ASM-based testing: Coverage criteria and automatic test sequence generation. *Journal of Universal Computer Science*, **7**(11):1050–1067, November 2001.

[23] A. Gargantini, E. Riccobene, and P. Scandurra. Model-driven language engineering: The ASMETA case study. In *International Conference on Software Engineering Advances, ICSEA*, pages 373–378, 2008.

[24] Angelo Gargantini and Constance Heitmeyer. Using model checking to generate tests from requirements specifications. In Oscar Nierstrasz and Michel Lemoine, editors, *Proceedings of the 7th European Engineering Conference and the 7th ACM SIGSOFT Symposium on the Foundations of Software Engeneering*, volume 1687 of *LNCS*, pages 6–10, Sep 1999.

[25] Angelo Gargantini, Elvinia Riccobene, and Salvatore Rinzivillo. Using Spin to generate tests from ASM specifications. In Egon Börger, Angelo Gargantini, and Elvinia Riccobene, editors, *Abstract State Machines, 10th International Workshop, ASM 2003*, pages 263–277, March 3-7 2003.

[26] Grégoire Hamon, Leonardo Mendonça de Moura, and John M. Rushby. Generating efficient test sets with a model checker. In *2nd International Conference on Software Engineering and Formal Methods (SEFM 2004), 28-30 September 2004, Beijing, China*, pages 261–270, 2004.

[27] M. Jean Harrold, Rajiv Gupta, and Mary Lou Soffa. A Methodology for Controlling the Size of a Test Suite. *ACM Transactions on Software Engineering and Methodology*, **2**(3):270–285, 1993.

[28] Mats P.E. Heimdahl and Devaraj George. Test-suite reduction for model based tests: Effects on test quality and implications for testing. In *Automated Software Engineering*, Linz, Austria, September 2004.

[29] R. M. Hierons. Comparing test sets and criteria in the presence of test hypotheses and fault domains. *ACM Trans. Softw. Eng. Methodol.*, **11**(4):427–448, 2002.

[30] G. J. Holzmann. The model checker SPIN. *IEEE Transactions on Software Engineering*, **23**(5):279–295, May 1997.

[31] Hyoung Seok Hong, Sung Deok Cha, Insup Lee, Oleg Sokolsky, and Hasan Ural. Data flow testing as model checking. In *ICSE'03, Portland, Oregon, May 3-10*, 2003.

[32] Hyoung Seok Hong, Insup Lee, Oleg Sokolsky, and Sung Deok Cha. Automatic test generation from statecharts using model checking. In *Proceedings of FATES'01, Workshop on Formal Approaches to Testing of Software, August 2001. BRICS Notes Series, NS-01-4, pp. 15-30.*, 2001.

[33] William E. Howden. Weak mutation testing and completeness of test sets. *IEEE Transactions on Software Engineering*, **8**(4):371–379, July 1982.

[34] IEEE. *IEEE Standard Glossary of Software Engineering Terminology*. Institute of Electrical and Electronics Engineers, **610**.12.

[35] G. Kaminski and P. Ammann. Using a Fault Hierarchy to Improve the Efficiency of DNF Logic Mutation Testing. In *ICST'09: Proceedings of the 2nd International Conference on Software Testing Verification and Validation*, pages 386–395, Washington, DC, USA, April 1–4, 2009. IEEE Computer Society.

[36] Kalpesh Kapoor and Jonathan P. Bowen. Ordering mutants to minimise test effort in mutation testing. In *Formal Approaches to Software Testing, 4th International Workshop, FATES*, pages 195–209, 2004.

[37] Kalpesh Kapoor and Jonathan P. Bowen. A formal analysis of MCDC and RCDC test criteria. *Softw. Test. Verif. Reliab.*, **15**(1):21–40, 2005.

[38] Y.G. Kim, H.S. Hong, S.M. Cho, D.H. Bae, and S.D. Cha. Test cases generation from UML state diagrams. *IEE Proceedings - Software*, **146**(4):187–192, Aug 1999.

[39] D. Richard Kuhn. Fault classes and error detection capability of specification-based testing. *ACM Transactions on Software Engineering and Methodology*, **8**(4):411–424, October 1999.

[40] D. Richard Kuhn, Dolores R. Wallace, and Albert M. Gallo. Software fault interactions and implications for software testing. *IEEE Trans. Software Eng*, **30**(6):418–421, 2004.

[41] Man Fai Lau and Yuen-Tak Yu. An extended fault class hierarchy for specification-based testing. *ACM Trans. Softw. Eng. Methodol*, **14**(3):247–276, 2005.

[42] D. Lee and M. Yannakakis. Principles and methods of testing finite state machines - A survey. In *Proceedings of The IEEE*, pages 1090–1123, August 1996. Published as Proceedings of The IEEE, volume 84, number 8.

[43] A. Jefferson Offutt and Stephen D. Lee. An empirical evaluation of weak mutation. *IEEE Transactions on Software Engineering*, **20**(5):337–344, May 1994.

[44] Vadim Okun, Paul E. Black, and Yaacov Yesha. Comparison of fault classes in specification-based testing. *Information and Software Technology*, **46**:525–533, 2004.

[45] Alexander Pretschner. Model-based testing in practice. In John Fitzgerald, Ian J. Hayes, and Andrzej Tarlecki, editors, *FM*, volume 3582 of *Lecture Notes in Computer Science*, pages 537–541. Springer, 2005.

[46] Sanjai Rayadurgam and Mats P.E. Heimdahl. Generating MC/DC adequate test sequences through model checking. In *28th Annual NASA Goddard Software Engineering Workshop (SEW 03)*, 2003.

[47] J. Schimd. Executing ASM specifications with AsmGofer. http://www.tydo.de/AsmGofer.

[48] Tatsuhiro Tsuchiya and Tohru Kikuno. On fault classes and error detection capability of specification-based testing. *ACM Trans. Softw. Eng. Methodol.*, **11**(1):58–62, 2002.

[49] Elaine Weyuker, Tarak Goradia, and Ashutosh Singh. Automatically generating test data from a Boolean specification. *IEEE Transactions on Software Engineering*, **20**(5):353–363, May 1994.

[50] Cemal Yilmaz, Myra B. Cohen, and Adam A. Porter. Covering arrays for efficient fault characterization in complex configuration spaces. *IEEE Trans. Software Eng*, **32**(1):20–34, 2006.

[51] Yuen Tak Yu, Man Fai Lau, and Tsong Yueh Chen. Automatic generation of test cases from boolean specifications using the MUMCUT strategy. *Journal of Systems and Software*, **79**(6):820–840, 2006.

[52] H. Zhu, P. Hall, and J. May. Software unit test coverage and adequacy. *ACM Computing Surveys*, **29**(4):366–427, December 1997.

In: Fault Detection: Theory, Methods and Systems ISBN: 978-1-61728-291-1
Editor: Léa M. Simon, pp. 253-295

Chapter 6

VIBRATION MEASUREMENTS AND ANALYSIS FOR MECHANICAL FAULT DETECTION IN PRODUCTION LINE

Cristina Cristalli**, Enrico Concettoni and Barbara Torcianti
Loccioni Group, Via Fiume 16, 60030 Angeli di Rosora, Italy

Abstract

Electromechanical components manufacturing companies more often ask for automatic on-line inspection systems in order to accurately monitor the characteristics of all their products and components.

Condition monitoring of manufacturing appliances is often based on the analysis of machines' vibrations, as the emergence of the fault can be indicated by the variation of the vibration signals that they produce. This is because when a machine or a structural component is in good condition, its vibration profile has the "normal" characteristic shape, and it will change as a fault begins to develop.

Piezo-electric accelerometers and microphones are the most common vibration sensors used in fault diagnostics. However, they appear to be less attractive for on-line quality control because of several limitations, such as invasiveness, problematic installation and high sensitivity to background noise.

For these reasons, the use of Laser Doppler Velocimeters (LDV) has become an increasingly popular technique owing to the non-contact principle of the laser.

* E-mail address: c.cristalli@loccioni.com; Tel. +39 0731 8161, Fax +39 0731 814700.

However, despite the advantages of LDV, vibration measurements on rough surfaces can be distorted by undesired surface effects, such as the speckle noise.

Acquired vibration signals have to be processed and analysed in order to identify the characteristics that allow distinguishing between the good and the faulty machinery. To this end, several techniques of signal processing have been developed in the last years such as FFT based analysis, Wavelet analysis, Order analysis, etc.

After a brief introduction on the vibration theory for on-line diagnosis, this chapter presents an overview of the most common vibration measurement transducers and data analysis techniques, followed by some examples of their applications in the development of test benches for industrial production lines both on electro-mechanical components, such as universal motors and on assembled complex systems, such as washing machines.

Introduction

It is well known that vibration tests allow the discrimination between good and faulty products and hence the analysis of vibration signals can be used for quality control on production lines [1], [2]. In fact, a variety of manufacturing defects normally show a variance in noise and vibration. In particular, electro-mechanical systems are often composed of a large number of moving parts, and their relative motions and dynamic interactions (frictions, shocks, etc.) produce vibrations that can be measured on the machines or on the structures connected to it. As different parts of the machine will vibrate with various frequencies and amplitudes, the measurement and the monitoring of vibration can be used to check the right functioning of the system and also to detect possible deterioration or damage in its components. Even though manufacturing has benefited from recent advantages in automation and robotics, evaluation of the emitted noise and vibration still largely relies on human operators and expertise. This is partly due to the nature of the problem which is quite complex and requires the combination of advanced sensor technology and highly sophisticated data analysis techniques. A limited available testing time and noisy environmental conditions in the production line aggravate the examination.

The aim of the manufacturers is to reduce the number of faulty samples at the end of the production line and the demands of standards in the final products are very high. Statistical tests on a random selection of samples cannot guarantee the quality check of 100 % of the production and therefore cannot be considered as an appropriate method to assure a high standard of quality. Thus, a complete and accurate test on the final products is an essential requirement to obtain the necessary information that permits an adequate control of the production process.

Anyway, despite its advantages, the vibration test is still one of the most complex and difficult tests to be implemented on the manufacturing line, due to a number of reasons:

- intrinsic complexity of vibration phenomena in assembled systems
- measuring conditions of the line
- cycle time available.

An automatic system should satisfy the following requirements:

- repeatability
- reliability
- flexibility
- capacity to be integrated into automatic production lines while other tests are being executed.

Nowadays, thanks to more reliable and selective measurement systems and Artificial Intelligence development, the previous points can be often reached and automatic quality control systems are becoming part of standard production lines.

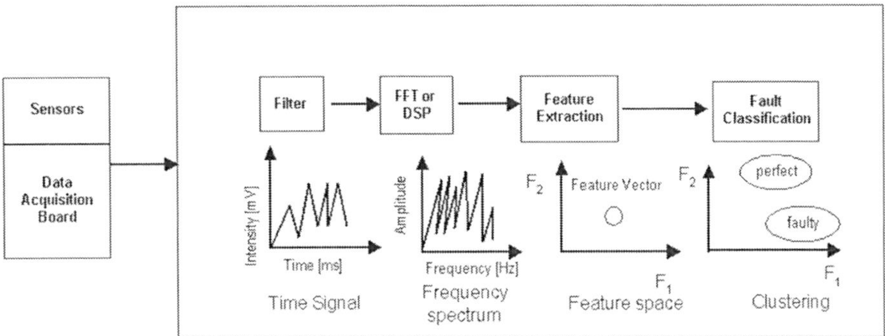

Figure 1. General approach in the vibro-acoustic field of applications.

In the following paragraphs, the fundamental steps in the design of an automatic vibration test station will be illustrated (Fig.1). In particular, the basics of vibration of mechanical systems and of signal digital acquisition will be reported and discussed as well as the selection of the appropriate transducer. In fact, vibration has to be measured by sensors that produce an electrical signal proportional to the mechanical vibration that they are able to detect as acceleration, velocity, pressure, etc. Then, these signals have to be recorded in a digital way, so that a computer-based analysis is made possible through advanced signal process techniques that are usually integrated in order to allow the feature extraction and the final classification.

An overview of the main applications in production lines of electrical motors and washing machines will be further discussed.

Fundamentals of Vibration and Data Acquisition Theory

The Vibration of Mechanical Systems

According to its definition, "vibration" is a mechanical oscillation around a reference position.

The easiest kind of vibrating system is the free un-damped mass-spring system. When the mass is moved from its equilibrium position and then released, it starts to oscillate in a sinusoidal way around the equilibrium position at the "natural frequency" of the system. This frequency can be easily calculated considering that as there is no energy dissipation, the kinetic energy of the mass must be always equal to the potential energy stored in the spring.

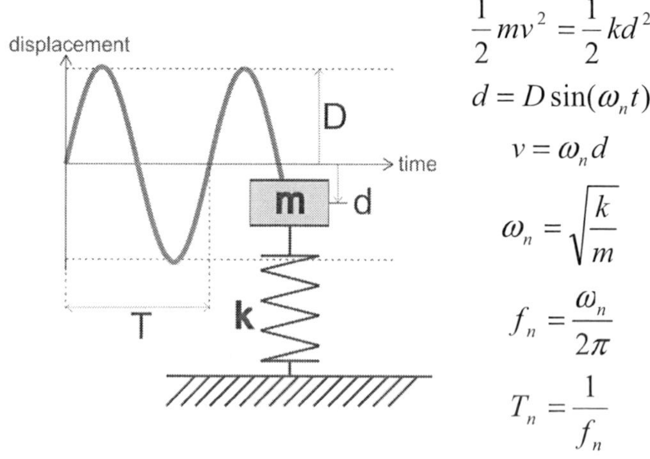

$$\frac{1}{2}mv^2 = \frac{1}{2}kd^2$$

$$d = D\sin(\omega_n t)$$

$$v = \omega_n d$$

$$\omega_n = \sqrt{\frac{k}{m}}$$

$$f_n = \frac{\omega_n}{2\pi}$$

$$T_n = \frac{1}{f_n}$$

Figure 2. Free un-damped mass-spring system and motion equations [1].

It is easy to observe from the previous equations that "rigid" systems, i.e. with high spring stiffness, have high resonance frequencies, while large structures, i.e. constituted by large masses, have low resonance frequencies.

When a linear damper is added, it absorbs energy proportionally to the velocity of the mass. In this case, a mathematical solution of the problem is possible using differential linear equation method (see [1] for further details). Once that the system is perturbed, it will start to vibrate at its damped natural frequency, that is constant and almost the same as the natural frequency. However, the amplitude of the vibration decreases during time: the higher is the damper, the faster will be the decay of the vibration.

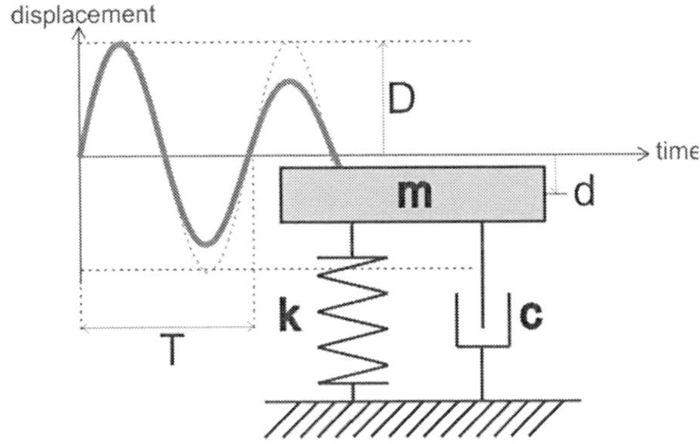

Figure 3. Free damped mass-spring system.

The system in Figure 3 is also called "single degree of freedom (SDOF) linear system", as its motion can be described using a single 2nd order linear differential equation.

When a SDOF is excited by a sinusoidal force, it follows the force, i.e. it moves at the same frequency of the external force, but the amplitude of the motion (and also the phase) depends on the frequency: at low frequencies, the displacement of the mass and of the one produced by the force on the other side are the same, then the amplitude of the vibrating system increases as the frequency is increased, and reaches its maximum at the natural frequency. If there were no damping in the system ($c = 0$), the amplitude will approach infinity. If the frequency of the external force is increased, the amplitude decreases and at very high frequencies it becomes close to zero.

Electro-mechanical systems generally consist of a large number of interacting masses, springs and dampers which can move in more than one direction; this means that they can be represented as a multi degree of freedom system (MDOF).

In this case the frequency spectrum has one peak for each degree of freedom, although it can often be difficult to separate the contribution of different mechanical components.

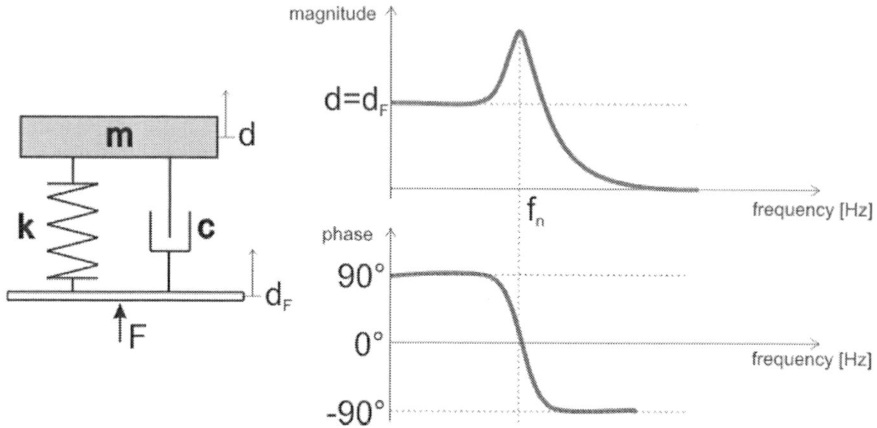

Figure 4. Forced SDOF system and frequency response functions.

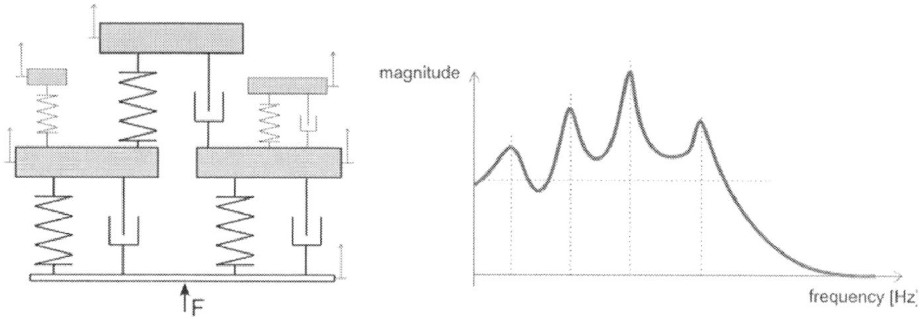

Figure 5. Forced MDOF system and frequency response function (magnitude).

When measuring the dynamic behaviour of a complex system, the vibration measured depends on the structure and on the characteristics of its components, but also on the nature of the exciting forces. In our case, i.e. the electro-mechanical systems diagnosis, forces are produced by the motion of internal components, as motors, shafts, power transmissions, connection rods, etc... Different kinds of motion produce a different kind of excitation (see Figure 6): for example a rotating component (shafts, fans, turbines, etc...) produces because of its unbalancing a sinusoidal excitation at its rotating frequency, while hits between parts (alternative motion, motion on rough surfaces, etc…) produce shocks that excite a broad band of frequencies.

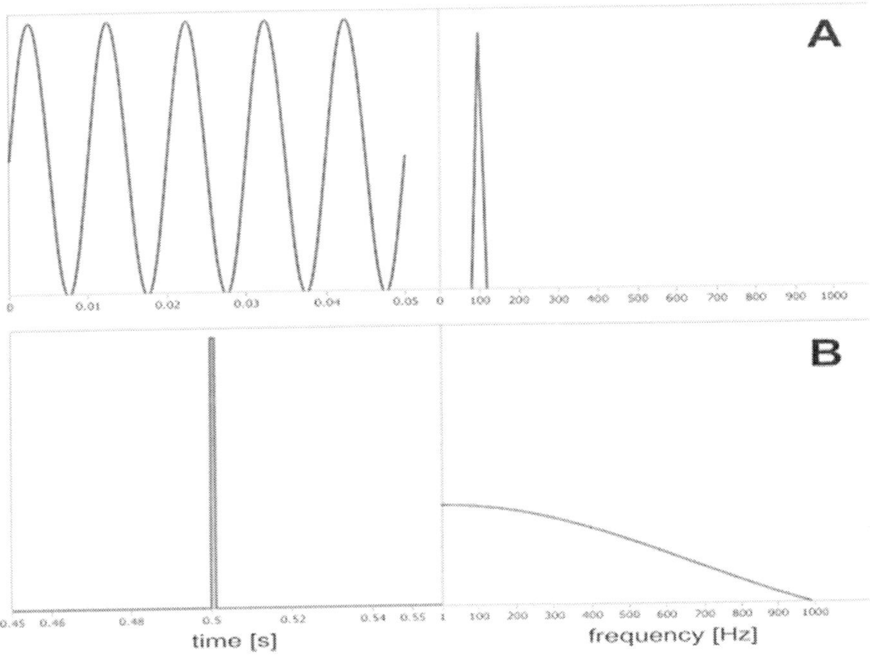

Figure 6. Time domain and frequency domain representation of a 100 Hz sine wave (A) and of a 1 ms pulse (B).

Data Sampling and Digital Data Analysis Fundamentals

When acquiring a signal from a vibration transducer with a digital data logger, there are some aspects that have to be taken in account in order to make a good data sampling, i.e. fully representative of the actual analog signal and without artefacts.

The first important thing to set when acquiring a signal is the sample rate (sr), i.e. the frequency at which data are collected. Looking at the figure above the sample rate is defined as:

$$sr = \frac{1}{dt}$$

where dt is the distance between samples along the time axis. The sr is very important because of the Nyquist–Shannon sampling theorem, which states that an analog signal is completely described only if

$$sr > 2f_{\max}$$

where f_{max} is the highest frequency in the signal to be acquired. The quantity $2f_{max}$ is often called Nyquist frequency. If sampling a signal with frequency components higher than sr/2, then "aliasing" artefact occurs: these components are still present in the signal but they are shifted at lower frequencies [6].

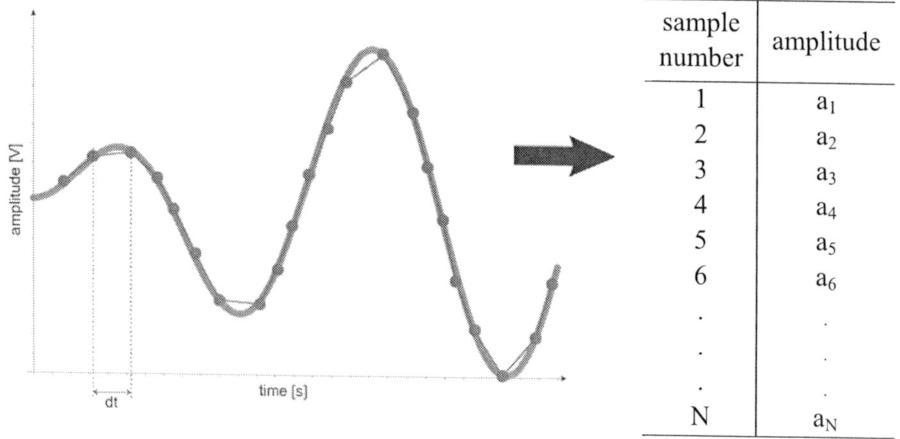

sample number	amplitude
1	a_1
2	a_2
3	a_3
4	a_4
5	a_5
6	a_6
.	.
.	.
.	.
N	a_N

Figure 7. A/D conversion of a signal.

The signal has to be sampled not only in the time axis, but also in the amplitude axis. Acquisition devices always have a stated number of bits (b), that is related to the maximum number of digital levels that it can be provided by the following law:

$$DL = 2^b$$

and DL is directly correlated to the resolution of the system. The minimum variation of the analog input signal that can be detected by the acquisition device is:

$$\Delta a_{\min} = \frac{acq.range}{DL} = \frac{acq.range}{2^b}$$

where *acq. range* is the amplitude range of the acquired signal without getting an overload (it can be usually set by the user).

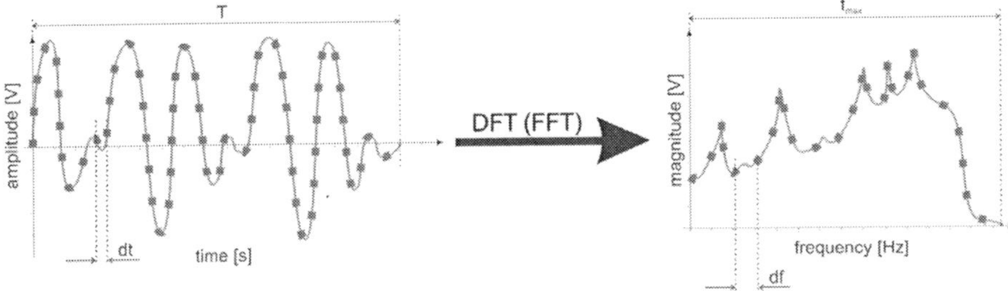

Figure 8. DFT and FFT.

Digital signals can be also represented in the frequency domain. The algorithms used to transform the signal from time to frequency domain are the Discrete Fourier Transform (DFT) or the Fast Fourier Transform (FFT) that can be used when the total number of samples acquired (N) is a power of two. Correlation between time domain sampling quantities and the analog ones in the frequency domain can be studied considering that the highest frequency represented in the spectrum is the Nyquist frequency and that the DFT/FFT algorithm uses half of the samples acquired to build the magnitude spectrum and the other half for the phase.

$$sr = \frac{1}{dt} \qquad T = n \cdot dt \qquad f_{\max} = \frac{sr}{2} = \frac{1}{2dt} \qquad df = \frac{f_{\max}}{N/2} = \frac{sr/2}{N/2} = \frac{1}{N \cdot dt} = \frac{1}{T}$$

Measurement Set-up

Figure 9 shows an example of a typical vibration measurement set-up.

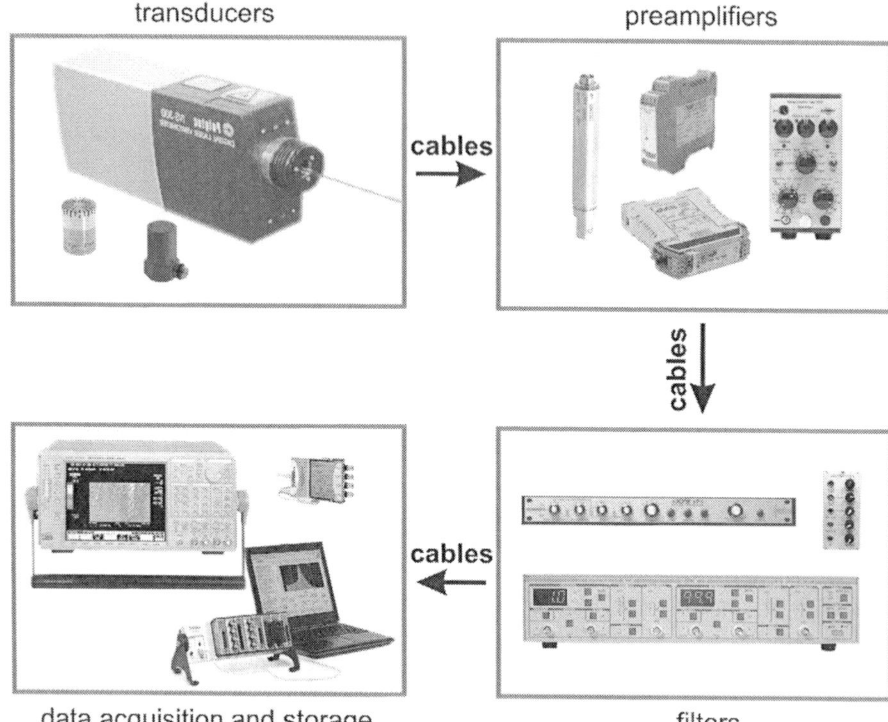

Figure 9. Example of vibration measurement chain.

In this paragraph a short overview of the most common sensors and devices for signal conditioning and data acquisition used for vibration measurement will be presented.

Sensors

Accelerometers

As the name says, accelerometers are acceleration transducers. The most common method used to manufacture accelerometers is based upon the principle of piezoelectricity: when a crystal of a piezoelectric (usually quartz or piezo ceramics) is loaded with a force, it exhibits electrical charges that can be collected putting two electrodes on the opposite sides of the crystal [2].

Figure 10 shows a cross-section of a typical piezoelectric accelerometer. The force is provided to the crystal by the inertia of a seismic mass which is mounted on the crystal through a stud that preloads the crystal, in order to measure acceleration in both directions.

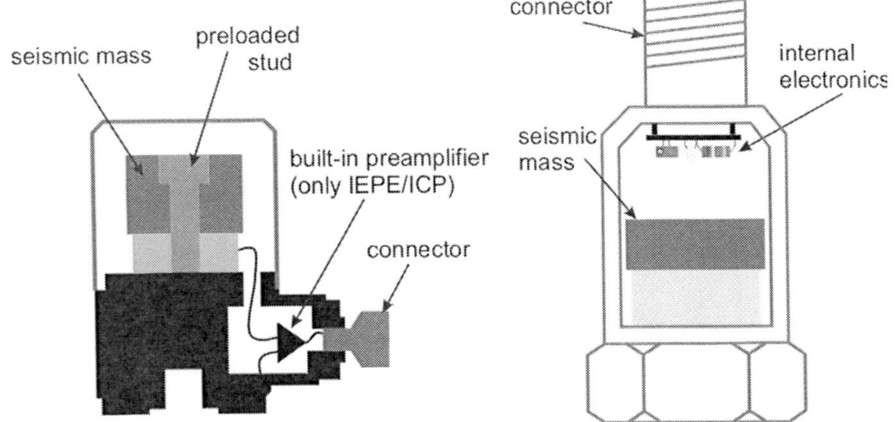

Figure 10. Piezoelectric and piezoresistive accelerometers.

Sensitivity is proportional to the size of the mass and of the crystal. The traditional construction gives a moderately high sensitivity/mass ratio but ensures a very stable design and a rigid structure; therefore it is especially used for high frequency or shock accelerometers.

To increase sensitivity two or more piezo crystals can be connected together for output multiplication, moreover other designs like shear or bending/deflection accelerometers can be adopted.

Piezoelectric accelerometers are the most common vibration sensors as they have a very large dynamic range and a wide frequency range, and in addition, they are compact, often with low weight and very stable.

To acquire the signal produced by the accelerometer's crystal it is necessary to use a charge-to-voltage converter. Traditional accelerometers use an external preamplifier but there are also accelerometers with a built preamplifier that only need an external DC constant current power supply (see section 2.2) and they are usually identified as IEPE (Integrated Electronics Piezo Electric) accelerometers, low-impedance accelerometers or ICP® (Integrated Circuit Piezoelectric) accelerometers (ICP® is the trademarked PCB name for IEPE accelerometers and is a property of PCB Group, inc., Depew, New York)

Traditional accelerometers are in general more versatile than the others, as gain and other parameters are adjustable from the external preamplifier and having no built-in electronics; they have a wider operating temperature range. For applications with a defined vibration range, however, internally pre-amplified accelerometers can cost less and furthermore can be used with general-purpose and longer cables.

Both types of accelerometers are AC-coupled systems, with limited low-frequency response that makes impossible to measure DC accelerations (gravity for example). That is because the piezoelectric crystal acts as a capacitor put in series with the circuit that stops DC and quasi-static current.

To measure DC or very low frequency accelerations it is possible to use piezoresistive accelerometers, which use a piezoresistive material in place of the piezoelectric crystal that changes its resistance according to the force exerted by the seismic mass (see Figure 10). Piezoresistive elements are usually connected in a Wheatstone bridge that transforms the resistance variation into a voltage signal.

To increase the sensitivity it is possible also in this case to adopt a bending/deflection design. Compared to piezoelectric ones, piezoresistive accelerometers have a smaller dynamic and frequency range and are also less stable.

Microphones

When an object vibrates in the presence of air, the air molecules at the surface will begin to vibrate, which in turn vibrates the adjacent molecules next to them. This vibration will travel through the air as oscillating pressure at frequencies and amplitudes determined by the original sound source [3]. As the human eardrum transfers these pressure oscillations into electrical signals that are interpreted by our brain, microphones are designed to transform pressure oscillations into electrical signals, which can be acquired in order to study the original source of vibration. Like the human ear, frequency range of microphones is typically 20-20000 Hz.

The most common designs are condenser microphones (externally polarized or pre-polarized), magnetic microphones and piezoelectric microphones.

A condenser microphone operates on a capacitive design. The cartridge transforms the sound pressure into capacitance variations which are then converted to an electrical voltage. This is accomplished by taking a small thin diaphragm and stretching it of a small distance away from a stationary metal plate, called a "backplate." A voltage is applied to the "backplate" to form a capacitor. In the presence of oscillating pressure, the diaphragm will move changing the gap between the diaphragm and the "backplate" and this produces an oscillating voltage from the capacitor, proportional to the original pressure oscillation.

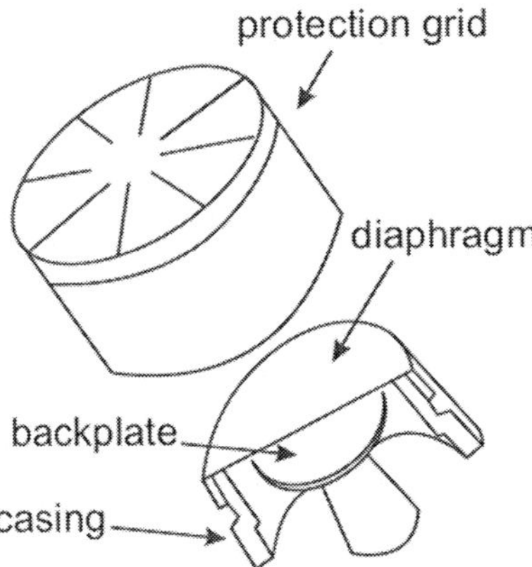

Figure 11. Capacitive microphone components.

If the "backplate" voltage is generated by an external power supply, the microphone is indicated as externally polarized microphones. This design can be coupled with an IEPE/ICP® circuit and can provide great advantages as they are less expensive and there is the possibility to use long coaxial cable without a degradation of the signal.

A magnetic microphone is a dynamic microphone. The moving coil design is based on the principle of magnetic induction. This design can be simply achieved by attaching a coil of wire to a light diaphragm. Upon seeing the acoustical pressure, the coil will move. When the wire is subjected to the magnetic field, the movement of the coil in the magnetic field creates a voltage which is proportional to the pressure exerted on it.

A piezoelectric microphone uses a quartz or ceramic crystal structure. Although these sensor type microphones have very low sensitivity if compared to the others, they are very durable and are able to measure very high amplitude pressure ranges, and this makes them suitable for shock and blast pressure measurements. Conversely, the floor noise level on this type of microphone is generally very high.

Laser Doppler Vibrometers (LDV)

LDV is based on the Doppler frequency shift of a coherent laser light, when it is back scattered from a moving object. The object scatters or reflects light from the laser beam and the Doppler frequency shift is used to measure the component of velocity which lies along the axis of the laser beam [4].

As a direct demodulation of the light is not possible (frequency of light is too high), an optical interferometer is therefore used to mix the scattered light coherently with a reference beam. The photo detector measures the intensity of the mixed light whose beat frequency is equal to the difference frequency between the reference and the measurement beam.

The easiest arrangement of a LDV is based on the Michelson interferometer as is shown in Figure 12.

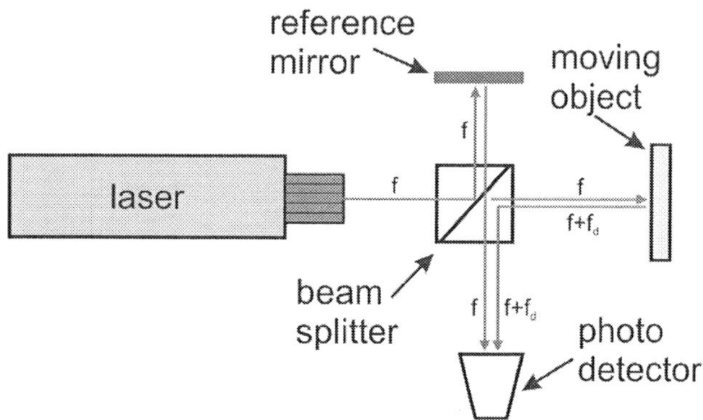

Figure 12. Michelson optical interferometer.

The laser beam emitted at a certain frequency f is divided at the beam splitter into a reference beam which propagates in the arms of the interferometer and a measurement beam, which is then backscattered by the surface of the moving object at a frequency $f+f_d$, where f_d is the Doppler shift. When the reference and measurement beams are joined together, they interfere, building a fringe pattern that the photo detector converts in a modulated voltage signal. From the analysis of this signal it is possible to calculate the velocity of the moving object (for further details see [5]).

The LDV discussed above is only able to measure the "out-of-plane" velocity of the measuring surface, but with the same principle and different laser/interferometer arrangement it is also possible to build "in-plane", rotational and differential vibrometers.

Thanks to the developing of solid state lasers and digital electronics, it has been possible to build up a very compact and robust LDV, suitable for measurement in severe working conditions as the production lines.

Their employment for mechanical systems diagnostics is rapidly increasing, as they allow a non-contact measurement and consequently to avoid the problem of positioning and fastening of the accelerometers.

Sensor Comparison for On-line Quality Control Test

Accelerometers are certainly the most used transducers for vibration measurements: they are produced in several models that cover almost all the requirements of the measurements for mechanical system diagnostic. Accelerometers are available almost in any kind of size and sensitivity: their dynamic ranges can vary from 50 m/s^2 for high sensitivity/resolution accelerometers to 10^6 m/s^2 for shock accelerometers, while their measuring frequency range can reach 30 kHz. Accelerometers' main disadvantage is the necessity to position and fasten them on the structure which makes them difficult to use in automated measuring system to integrate in production lines. Moreover, the load effect of their mass may not be negligible when the tested component is small.

Microphones are also widely used: they allow a non-contact measurement and are also adaptable to different measurement needs. Common microphones have a frequency range from about 3 Hz to over 20 kHz, but there are also available microphones that can measure over 100 kHz. The main difficulty of using microphones in production line vibrational tests is that they are sensitive to the very high background noise coming from other moving machineries and that it is often louder than the noise produced by the system under test during the functioning. In this case, measurement is still possible, but the microphone has to be positioned very close (few centimetres) from the component under test.

LDV in their "industrial" arrangement allow a non contact measure of vibration and this is certainly their major facility when compared with the other sensors. Moreover, in measuring the velocity, they are more suitable than piezoelectric accelerometers and microphones to measure very low frequencies, while their maximum measurement frequency can vary from 22 kHz for digital industrial vibrometers to 2-3 MHz for special vibrometers. Their main disadvantages are that the quality of the measure is strictly correlated to the amount and to the characteristics of light backscattered by the measured surface (for further details see [5]) and their high cost if compared to the cost of an accelerometer or a vibrometer.

Preamplifiers and Signal Conditioning

Charge Amplifiers

Charge amplifiers have to be used with piezoelectric sensors in order to convert the high impedance charge difference produced by the piezoelectric crystal into a usable output voltage.

In general they allow setting the gain of the signal but also the frequency range of the measure, as it is possible to modify the time constant of the AC-coupled circuit connected to the accelerometer.

IEPE/ICP® Conditioners

For excitation of their internal electronics, IEPE/ICP® accelerometers require a DC constant current power supply (generally 18-24V/2-4mA) that is connected to the accelerometer with only a single coaxial cable. Both the power into and the signal out of the sensor are transmitted over this cable. The coupler provides the constant current excitation required and also decouples the bias voltage from the output.

Filters

Analog filters are devices that can be very useful for signal conditioning before its acquisition. Passing over at all to their mathematical discussion, analog filters allow to remove unwanted frequencies from the original signal and they are characterised by one or two cut-off frequencies and by an "order", that determines its slope at the cut-off frequencies. They can be roughly classified in: low-pass, high-pass, band-pass and band-stop (see figure 13).

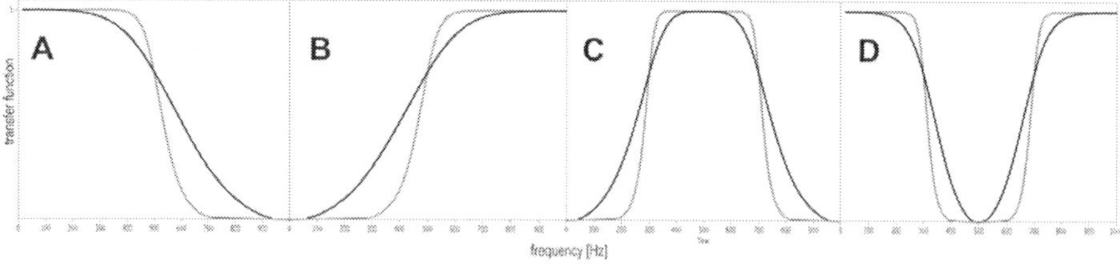

Figure 13. Common filters transfer functions: 500 Hz low-pass (A), 500 Hz high-pass (B), 300-700 Hz band-pass (C) and 300-700 Hz band-stop (D). Black line is a 2nd order filter, black line is a 6th order one.

As the name says, low-pass filters remove high frequencies and they can be very helpful to remove frequency noise from the signal and possible aliasing effect that can occur when sampling the signal. High pass filters are useful when low frequency artefact can occur in electrical signals, for example DC bias, zero drifts and piezoelectric accelerometers DC shift [7]. Band-pass and band-stop filters are the serial sum of a low-pass and a high-pass filters; they can be used to limit the frequency range of the signal or to remove specific frequencies where high noise occurs, as for example the mains electrical noise that often arises around 50-60 Hz.

Data Acquisition

This brief discussion on parameters of data acquisition devices is focused on "computer based" devices that have been used in the studies reported in Section 4 and represents the actual standard for automatic system implemented in production line test benches.

The choice of a Data Acquisition device (DAQ) is based on the analog input specifications such as: the number of channels, sampling rate, resolution, and input range [6].

The number of analog channel inputs is usually specified for both single-ended and differential inputs on boards that have both types of inputs. Generally it is preferable to set channels in differential mode, because, each input has its own ground reference and the noise errors are reduced because the common-mode noise picked up by the leads is cancelled out. In the single-ended configuration, all channels are referenced to a common ground point. This doubles the number of available inputs but signals acquired are usually more noisy and subjected to low frequency fluctuations. Anyway it can be used when the input signals have high level (greater than 1 V), the leads from the signal source to the analog input hardware are short (less than 5 m) and all input signals share a common ground reference.

The sampling rate must be chosen according to the Shannon–Nyquist theorem and depends on the maximum frequency of the original signal. A faster sampling rate acquires more points in a given time and can therefore often form a better representation of the original signal. DAQ boards can have one digital converter for each input channel or use a multiplexing technique. In the second case, the converter samples one channel, switches to the next channel, samples it, switches to the next channel, and so on. In this case the maximum rate of each individual channel is inversely proportional to the number of channels sampled.

The resolution is indicated by the number of bits that the digital converter uses to represent the analog signal. The higher is the resolution, the higher is the number of divisions the range is broken into, and therefore, the smaller will be the detectable voltage change (see Section 1.2).

The range refers to the minimum and maximum voltage levels that the converter can quantize. Usually multifunction DAQ boards offer selectable ranges and this allows matching the signal range in order to get the best resolution (see section 1.2).

Data Analysis

Once data are acquired, they have to be analysed in order to extract from them significant features for the system diagnostics. The analysis can be conducted in the time, in the frequency or in the joint time-frequency domain. The most common signal processing techniques employed for the system vibration analysis are reported as follows.

Time Domain

RMS

Root Mean Square (RMS) is the quadratic mean of a set of values and represents a statistical measure of the magnitude of a varying quantity. It is especially useful when periodic signals have a null mean.

For a continuously varying time function f(t), the RMS value in the interval $[t_1,t_2]$ can be calculated as:

$$f_{RMS} = \sqrt{\frac{1}{t_2 - t_1} \int_{t_1}^{t_2} [f(t)]^2 \, dt}$$

while for a discrete data set $x_1, x_2, \ldots x_N$:

$$x_{RMS} = \sqrt{\frac{1}{N} \sum_{i=1}^{N} [x(i)]^2}$$

Sometimes a simple measurement of the RMS value of the vibration signal may be enough to asses a fault of a system, as some kinds of damages (component missing or unscrewed, bad assembling, etc…) produce a general increase of the amount of vibration produced that makes the RMS value of the signal increase.

Envelope

Envelope detection is an amplitude-demodulation process in which all frequency components in a certain frequency range are demodulated. This technique has become one of the most important methods for diagnostics of rolling element bearings [8] and for detection of typical defects in induction motors, such as bad soldering and broken (or cracked) rotor parts [9]. In the following paragraph the principles of envelope detection will be rapidly illustrated; for further details please refer to [10] and [11].

The envelope detection is based on three steps: band-pass filtering, demodulation, low-pass filtering (see Figure 14).

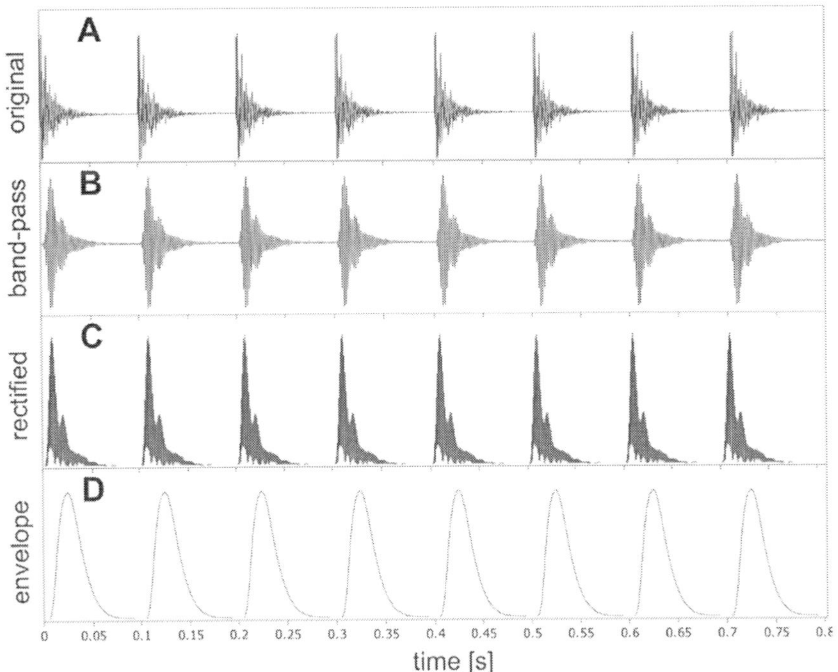

Figure 14. Example of an envelope detection of a typical vibration signal produced by a faulty bearing.

A first band-pass filtering is necessary to select the frequency band of interest avoiding to collect noise from other moving machinery in the line, then the signal has to be demodulated with a full wave rectifier if operating with analog devices or just calculating its absolute value if it has been already digitized. The rectified signal is then smoothed by low-pass filtering or a running average or RMS and this gives the envelope signal shown in Figure 14D.

Kurtosis

Kurtosis is a measure of the "peakedness" of the probability distribution of a variable. Higher kurtosis means that the most of the variance is due to infrequent extreme deviations, as opposed to frequent modestly-sized deviations. It is a useful quantity to individuate portions of the signal affected by spike shaped noise, as the speckle noise that can frequently occur measuring with laser Doppler vibrometers on not cooperative surfaces [5].

Kurtosis is commonly defined as the fourth standardized moment of the sample distribution and can be calculated as the 4th moment about the mean divided by the square of the variance of the probability distribution:

$$Kurt = \frac{\mu_4}{\sigma^4}$$

Sometimes a "minus 3" is added at the end of this formula in order to make the kurtosis of the normal distribution equal to zero.

For a sample of x_1, x_2, ... x_N values the sample kurtosis can be calculated with the following formula:

$$Kurt = \frac{N\left(\sum_{i=1}^{N}(x_i - \bar{x})^4\right)}{\left(\sum_{i=1}^{N}(x_i - \bar{x})^2\right)^2}$$

Frequency Domain

FFT Based Techniques

The analysis of spectra is the first and the most common method to analyse data of vibration measurements on mechanical systems.

The first and easiest method to assess the damage on a component or on an assembled machine when measuring in production line is to compare the vibration spectrum measured on the machine during the test with an average spectrum obtained from several machines classified as "good".

When damage is present, it can be recognised by the increasing of magnitude of some peaks or by the presence of peaks not present in the average good spectrum (see Figure 15).

Sometimes, instead of FFT (or DFT) spectrum it is preferable to use the Power Spectrum, which can be obtained squaring the FFT and can be useful to highlight peaks.

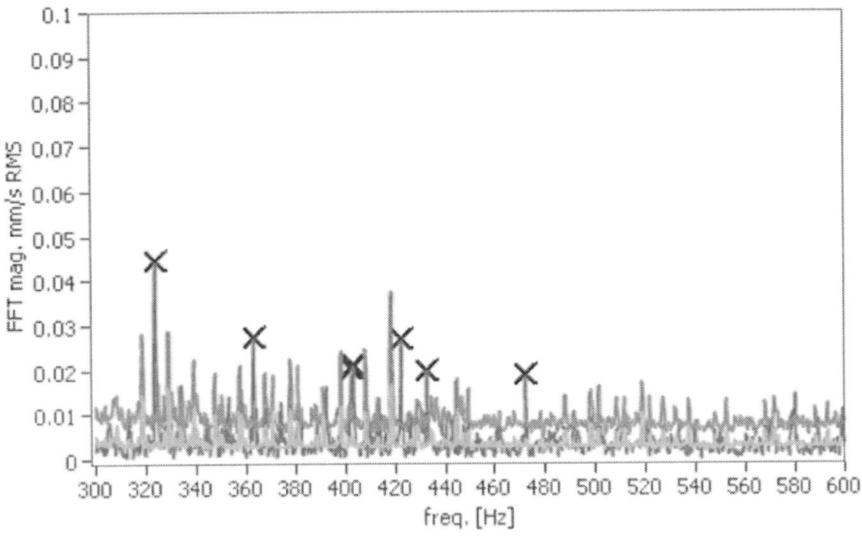

Figure 15. Example of an FFT spectrum analysis of a faulty bearing: green line is the average of a "good", red line is the upper boundary limit and the blue line is tested component.

FFT Enhancement: Cepstrum

The cepstrum is defined as the inverse Fourier transform of a logarithmic spectrum[1], and enhances periodic spectrum structures. It can be helpful when used together with the envelope spectrum that often contains harmonics of uniform spacing. The main advantage is that each whole family of harmonics is represented by a single line in the cepstrum (see Figure 16) and it can be very

[1] Some text define the cepstrum as "the Fourier transform of the logarithm (with unwrapped phase) of the Fourier transform of the signal", i.e. a spectrum of a spectrum.

useful to highlight families of harmonics generated by a rotating component damage [12] [13].

The name “cepstrum” was derived by reversing the first four letters of "spectrum"; in the same way the quantity corresponding to the frequency is the “quefrency”, which has the physical dimension of a time.

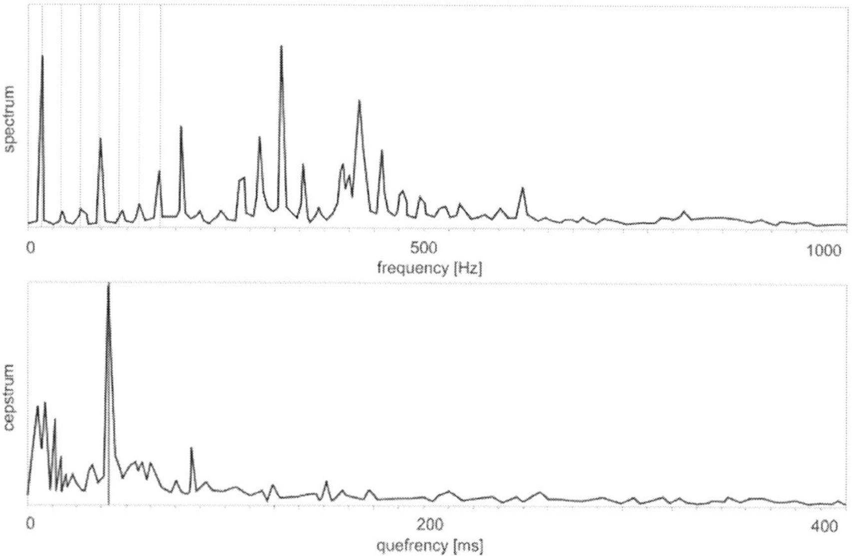

Figure 16. Example of how cepstrum acts: a spectrum characterize by the presence of many harmonics is a cepstrum with only one high peak.

It is possible to calculate a complex cepstrum using the inverse FFT of the complex logarithmic spectrum or a real cepstrum by determining the real logarithm of the amplitude spectrum, and then computing the IFFT of the resulting sequence. Similarly, the power cepstrum is obtained by the inverse FFT of the logarithmic power spectrum.

In formulas:

$$C(\tau) = \Im^{-1}\{\log F(f)\}$$

where $F(f)$ is a complex spectrum or direct Fourier transform.

Time-Frequency Domain

Short-Time Fourier Transform (STFT)

The Short-Time Fourier transform (STFT) is a Fourier-related transform used to determine the sinusoidal frequency and phase content of local sections of a signal as it changes over time [14].

It is obtained breaking up the acquired signal into chunks or frames, which usually overlap each other, and transforming each one with FFT. The complex result is added to a matrix which records magnitude and phase for each point in time and frequency.

The most usual way to represent the STFT is the "spectrogram", where the horizontal axis represents time, the vertical axis is frequency, and the colour represents the amplitude of a particular frequency at a particular time (see Figure 17).

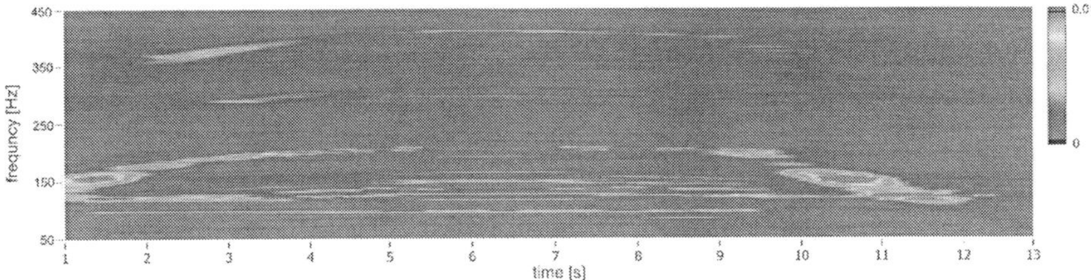

Figure 17. Example of a spectrogram measured on the cabinet of a washing machine with an accelerometer: horizontal lines represent structural resonance while the others follow the RPM of the motor.

Order Analysis

Order analysis is a vibration analysis technique, where multiples of machine speeds (orders) are used instead of absolute frequencies [15] [16] [17]. For rotating machinery, an order is a harmonic analysis [18]; since orders correspond to harmonics of the rotational speed (which itself represents the fundamental frequency). For this reason, orders are particularly useful for analysing speed-related vibrations, since often significant orders relate directly to various rotating parts of the machine [17], e.g. it has been observed [19] [20] that unbalance results in excitation of the first order, that misalignment or bending of the shaft generates a large second order and that gears, belts and blades might increase high orders.

Order tracking is a process of following (tracking) a specific order and possibly reconstructing its time-domain waveform and it is particularly important when the rotational speed varies, such as during the run-up and run-down of the machine.

The rotational frequency is commonly used for converting a frequency spectrum into an order spectrum, simply by normalising the frequency axis. Instantaneous rotational frequency can be measured directly with an encoder or other devices directly on the rotating shaft, or can be calculated from the same signal used for the vibration analysis, using frequency tracking algorithms in the time domain or in the frequency domain observing the phase of the signal at the first resonance peak that usually corresponds to the rotational speed.

Wavelet

Wavelet analysis [21] [22] overcomes the drawbacks of traditional FFT analysis and STFT processes.

Fourier Transform characterizes the signal in the frequency domain, but information about how the frequencies change over the time is completely lost. STFT simultaneously characterizes the signal in time and frequency domain, but with a fixed resolution once the type of window has been selected.

Wavelets are mathematical functions that cut up data into different frequency components, and then study each component with a resolution matched to its scale. They have advantages over traditional Fourier methods in analysing physical situations where signals have a long time period at low frequency and a short time period at high frequency, as the majority of signals encountered in nature.

Following a comparison of the three transform processes:

All the processes employ the same mathematical tool, i.e. the inner product, in order to compare the signal $s(t)$ with the elementary function, but the structure of the elementary functions $e_{\alpha}(t)$ is different.

In SFTF the elementary functions are time shifted and frequency-modulated single. The time and frequency resolutions of the elementary functions are constant, as the modulation does not change the time and frequency resolutions.

In Wavelet transform the elementary functions are dilated and shifted: by increasing m, the width of the wavelet is reduced, i.e. the time duration is compressed. The time resolution of the wavelets improves and the frequency resolution becomes worse as m increases. Due to this, wavelet analysis has a good time resolution at high frequencies and good frequency resolution at low frequencies. The parameter m can be considered as the *scale factor* and 2^{-m} as the *sampling step*. The shorter is the time duration, the smaller is the time sampling step, and vice versa.

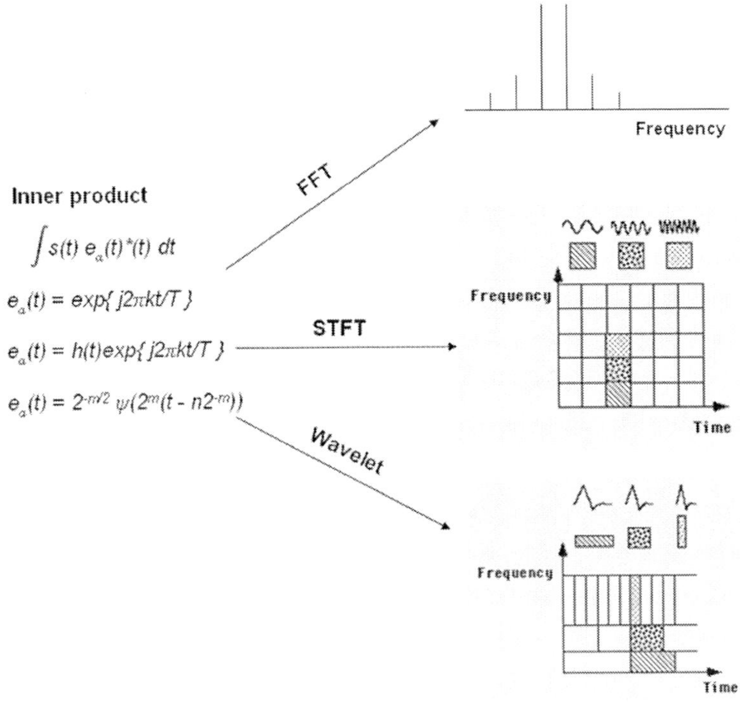

Figure 18. Schematic comparison of transform processes.

The *Continuous Wavelet Transform (CWT)* is generally represented as follows:

$$CWT_x(a,b) = \int_{-\infty}^{+\infty} x(t)\psi_{ab}^*(t)dt = < \psi_{ab}(t), x(t) >$$

where:

$$\psi_{ab}(t) = \frac{1}{\sqrt{|a|}}\psi\left(\frac{t-b}{a}\right)$$

$\psi(t)$ is named mother wavelet, a is the scaling parameter and b is the translation parameter.

The representation in figure 18 is a special case of a mother wavelet as following specified:

$\psi_{mn}(t) = \frac{1}{\sqrt{a}}\psi\left(\frac{t-b}{a}\right)\Big\vert_{a=2^m, b=2^m nb_0} =$	$2^{-\frac{m}{2}}\psi(2^{-m}t - nb_0)$

Wavelet methods provide powerful tools for analysing, compressing and reconstructing signals and images. They are useful in identifying multiscale and no stationary processes, including detecting the discontinuity of a signal, removing the trend, suppressing noise.

Examples of Vibration Measurement Applications for On-line Testing

Quality control of finished products is an essential part of the manufacturing process.

Statistical tests on a random selection of samples from the lines cannot guarantee the quality of the 100% of the production and therefore, they cannot be considered as appropriate methods to assure a high standard of quality control. Moreover in most cases, if 100% of on-line quality control is performed, it is carried out by an operator who listens to the machinery and audibly detects the existence of any malfunctions. The capacity of the operator to discriminate the different types of defects cannot be easily surpassed; however, one of the disadvantages lies on the lack of reproducibility due to the subjective component of the listener whose attention can vary along the working period. This is the reason why an automatic system that guarantees the control of 100% of production is increasingly required.

The reduction of the number of faulty samples by a pass or fail test at the end of the production line is the scope of many manufacturers who have adopted quality control as a key part of the production process. Classification of specific defects is also a relevant issue, because it allows feedback information to the assembly line to improve overall process control and efficiency, as well as to understand the causes of faults and take proper corrective actions.

The following part presents an overview of some vibration measurement systems and data analysis techniques applied for the quality control of electro-mechanical systems in the production line. In particular, applications in the development of test benches for industrial production lines of universal motors and washing machines will be presented.

Vibration Measurements on Electro-mechanical Components: Universal Motors

A test bench has been designed and realized with the aim of analysing the vibrations produced by universal motors of washing machines, so to define possible mechanical defects.

Instrumentation Set-up and Data Acquisition

In order to quantify the amplitude of the vibrations, the system uses two piezoelectric accelerometers applied to the external structure of the motor and two Laser Doppler Vibrometers (LDV), one pointing the upper bearing and the other one pointing the lower bearing (Figure 19). Such a system is able to detect problems related to the rotor, brushes, unbalance, bearings and cage of universal motors.

Figure 19. Measurement set-up for quality control of universal motors by accelerometers and LDV.

A non-contact measurement technique has been adopted in order to overcome problems related to sensor installation. In fact, the bearing housing is a small curved surface which does not allow the use of fixing devices such as magnets.

The LDVs are isolated from the external vibration by rubber elements and they are located at about 40 cm from the motor. The controller demodulates the vibration and provides an analog voltage proportional to the velocity of the surface.

The accelerometers are positioned on a support and connected to the stator by means of a magnet. Moreover they are managed by a mechanical hand which positions them on the magnetic structure of the motor when the test starts and removes them at the end of the test. In this case the surface allows the positioning of the transducers through a magnet.

The analog signals from the accelerometers and the vibrometers are then acquired by means of a 16 bit acquisition board. Data are sampled in the time domain with a sampling frequency of 50.000 Hz.

The vibration signals are analysed in order to reveal anomalous vibrations of the motors and to classify them as good or faulty. A LabVIEW® program, especially developed for this purpose, activates and controls the mechanical handling, the functioning of the motor, the acquisition board, and the signal processing.

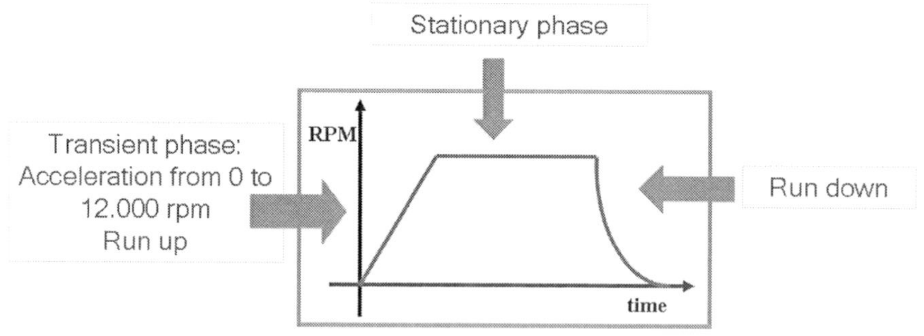

Figure 20. Acceleration curve of the motor.

During the measurements the motor is standing with its axis in the vertical position; moreover it is positioned on a pallet properly insulated from the assembly line. The measurement process consists basically in accelerating the motor, to keep it rotating at constant speed for several seconds and finally to stop it. The transient phase comprises the run up of the motor from 0 to a prefixed threshold of 12.000 rpm, when the motor has reached the maximum speed. This phase lasts around 4 s in a typical test. After this transient phase, the motor runs in stationary state for about 6 s, which means a total time of 10 s required to acquire data during both phases. Then the motor is stopped.

At the end of the test, the motor is properly marked according to the results of the vibration measurements.

Signal Processing

Despite the advantage of the LDV, speckle noise occurs when the measuring surface is rough, as in the case of bearing housing, or the object is moving. Therefore speckle noise removal [23] [24] [25] is a fundamental step to be performed before the automatic identification of faults.

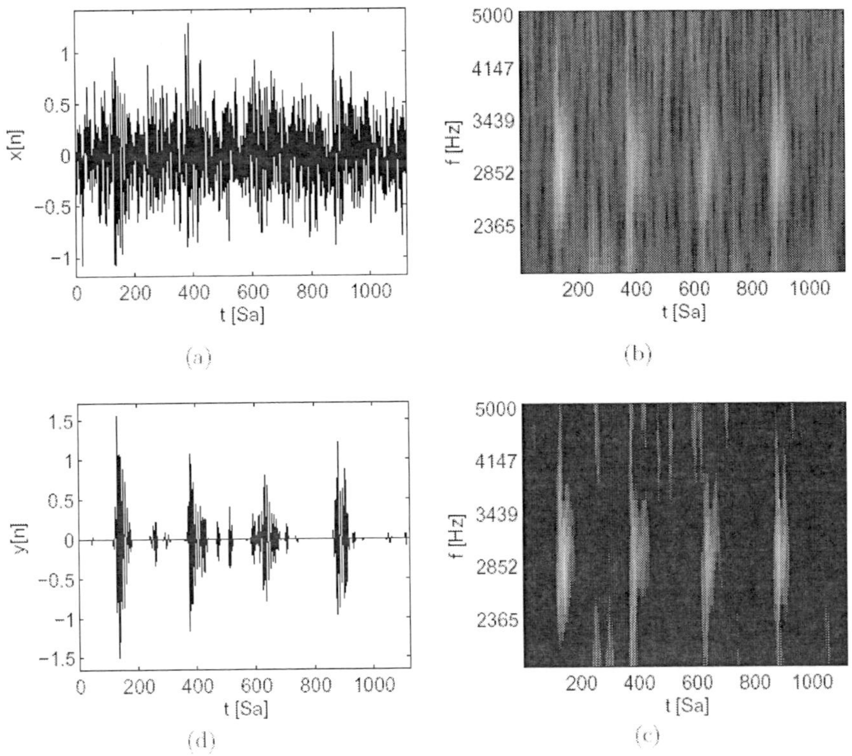

Figure 21. Denoising process through wavelets.

In order to reduce this undesired effect, a processing method based on CWT and noise suppression by thresholding of wavelet coefficients has been adopted.

The process can be easily described by the following four steps described in Figure 21:

- Figure 21a shows a vibration signal corrupted by noise: it refers to a faulty bearing, but the characteristic impulses related to the fault are completely buried in the noise and they are not visible;
- Figure 21b shows a *scalogram* of the noisy signal. Here periodic impulses can be seen;
- Figure 21c displays the scalogram after application of wavelet coefficient thresholding;
- Figure 21d shows the denoised signal reconstructed by the inverse CWT.

Then the signals are processed as follows:

- TRANSIENT STATE (1000 RPM -12000 RPM):
  - a STFT is performed on the signal acquired by one of the accelerometer;
- STEADY STATE: (12000 RPM):
  - 1/3 octave bands analysis is performed on the signals acquired by the accelerometers (axial and radial) and LDVs (upper and lower);
  - a FFT is also calculated on signals acquired by LDVs (upper and lower).

Feature Extraction and Classification

Both in the transient state and in the steady state characteristic features have been extracted in order to allow the classification of a motor as good or faulty.

In the TRANSIENT STATE, the features are related to the energy of two selected windows in the time-frequency domain.

The discrimination is then automatically carried out by establishing a threshold value for every feature. Every motor that surpasses that limit will be considered as faulty, as shown below:

In the STEADY STATE each graph of Figure 23 shows the spectrum of the signal represented by the values of the 18 bands of 1/3 of octave analysis (between 200 Hz and 10000 Hz) and also the upper limit of each band (red thin vertical line). If a red line invades the corresponding green bar it means that the band has exceeded the limit and this is signalled by the red indicator over the considered band (red colour → negative result, green colour→ positive result).

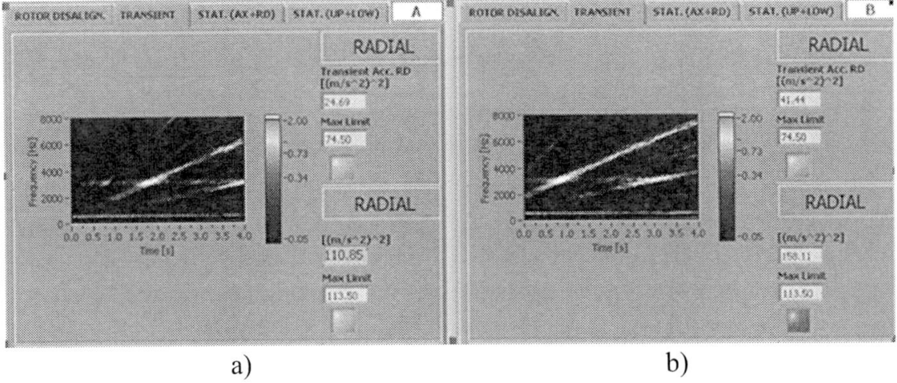

a) b)

Figure 22. STFT of a good (a) and faulty (b) motor.

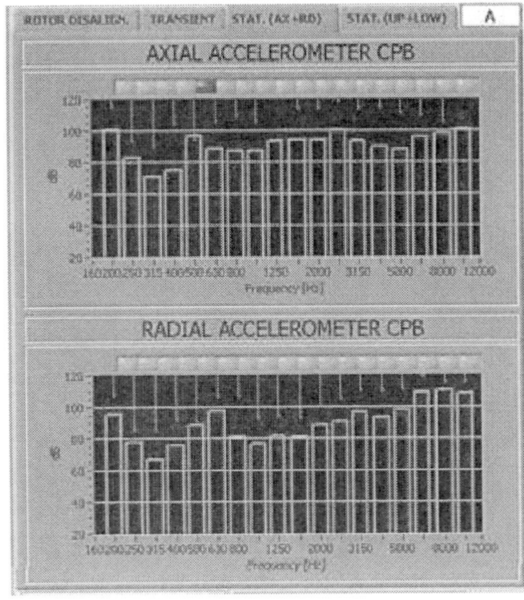

Figure 23. 1/3 octave band analysis for the axial and radial accelerometer.

The results referred to the axial and radial accelerometer are presented in fig.23.

Bands of 1/3 of octave analysis (between 200 Hz and 20000) are calculated also in the LDV signals; but, in this case, only two features per laser are extracted.

In addition, other two features have been extracted from the FFT analysis of the signals acquired by the LDV.

The spectra of a good and faulty motor are shown in fig. 24.

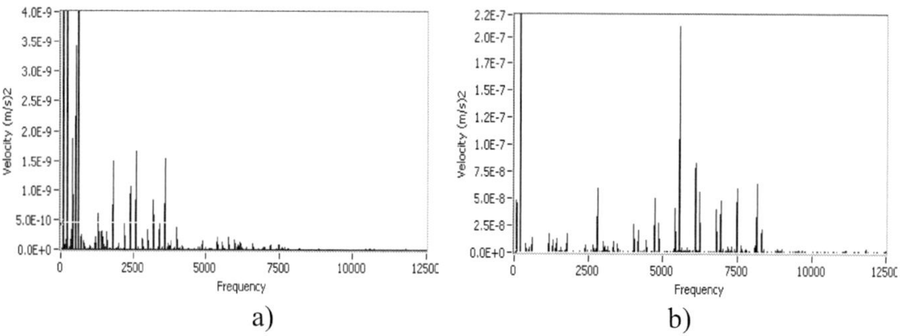

Figure 24. FFT analysis of velocity signal of a good motor (a) and a faulty motor (b) – signals acquired by the lower LDV.

In order to extract meaningful features from the signals acquired by the LDVs, instead of looking at frequency spectra in the frequency domain, it is also possible to look at the order domain [22]:

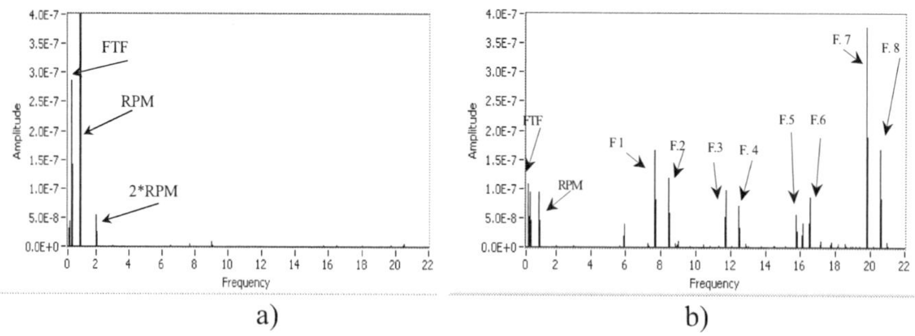

Figure 25. Velocity order spectrum of good motor (a) faulty motor (b) with defect in the lower bearing – signals acquired by the LDV pointing the lower bearing.

The order analysis has been used to detect the presence of these frequencies, always relating them to the rotational speed. In this particular case, according to the geometrical dimension and the type of bearing existing in the motor, the orders (characteristic frequencies divided by the rotational speed, 12.000 rpm) will be the following:

- FTF frequency represents the Fundamental Train Frequency and reveals the existence of defects in the bearing cage;
- RPS is the rotational speed of the motor;
- BSF is the Ball Spin Frequency and shows the defects in the rolling element;
- BPFO indicates the Ball Pass Frequency of the Outer Race;
- BPFI Orders represents the Ball Pass Frequency of the Inner Race.

Table 1. Order frequencies according to the type of bearing and defect

	Order (Frequency/RPS)	
Main frequencies	**Lower Bearing**	**Upper Bearing**
FTF	0.36	0.38
RPS	1	1
BSF	1.71	2.03
BPFO	2.55	3.07
BPFI	4.45	4.93

Both BPFO and BPFI reveal defects in the outer and inner ring respectively.

The orders of rotation can be considered as harmonics, but unlike harmonics, many interesting orders are not integer harmonics of the 1st order (of the rotational speed), as it can be seen in the table 1.

A general observation of the order spectra has shown that, within the range of the lower orders, the differences between a good and a faulty motor are not well distinguished. However, a large difference appears in the higher orders which reveal features quite sensitive to defects. Therefore the attention has been concentrated on these features which are discussed hereafter for one of the defects in the lower bearing as an example. Similar discussion could be made for other types of defects. The most relevant frequencies that can be observed in the spectra of the good motor are related to FTF, the rotational speed RPM and some harmonics; the spectrum of the faulty motor reveals components in the higher orders, which also exist in the good one, but with much lower amplitude. These components are combinations of the characteristic frequencies of Table 1 and a subset of them has been chosen. They have been defined as feature 1 (F1), feature 2 (F2), feature 3 (F3), etc.

Velocity spectra measured by laser vibrometer have been compared with these features in a series of good and faulty motors, in order to find correlation with the defect. In Figure 26 the amplitudes of the velocity power spectrum show that there is a clear difference between the amplitudes of good and faulty motors, which are grouped in two sets. Therefore this confirms that laser vibrometer data can be well correlated to the defect.

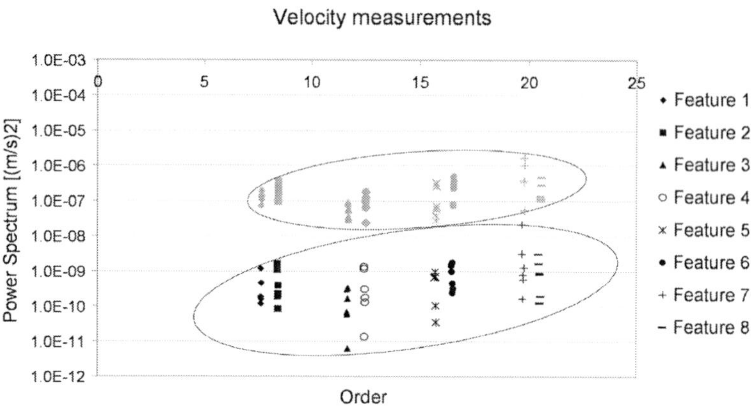

Figure 26. Comparison of velocity power spectrum amplitudes between good (black) and faulty motors (grey) with defect in the lower bearing - signals acquired by the LDV pointing the lower bearing.

Vibration Measurements on Assembled Household Appliances: Washing Machines

A washing machine (WM) is an electro-mechanical system, basically composed of a metal cabinet which hosts internally a tub which in turn contains the rotating drum, driven by an electric motor, in most cases through a pulley. The tub is suspended to the cabinet by vibration insulators, generally springs and dampers and an inertial mass. So, the WM has many electro-mechanical subsystems which may have faults or may be not properly assembled; several moving parts strongly affect the overall performance, both in terms of functionality and in terms of vibro-acoustic comfort. Therefore it represents a very challenging test case for Intelligent Data Analysis Systems.

Instrumentation Set-up and Data Acquisition

Figure 27 presents the on-line test station developed for the purpose of vibration-based quality control of WMs.

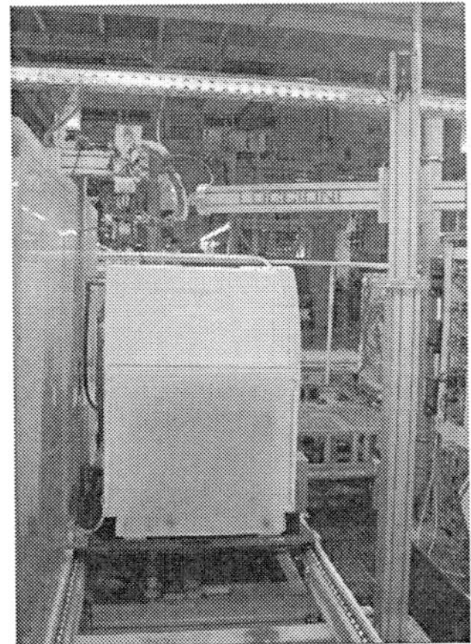

Figure 27. Measurement lay-out in production line.

The kernel of the overall system is constituted by the PC-unit. It controls all the operations for the handshake with the production line, the electronic feedback, the data acquisitions and computational processes necessary to the classification of the machine under test.

General faults of WMs are due to defects of its parts: screws, dampers, belt, counterweight, pulley, bearings. But the most frequent fault is related to the electrical motor due to the rotating parts, namely the bearings and the rotor. The diagnostic station should identify these faults, in particular the ones related to the motor.

As sensor one industrial LDV is used: it points the tub of the WM, in radial direction respect to the axis of the motor. The velocity signal is acquired together with the correspondent quality signal by a 16 bit data acquisition board with a sampling frequency of 20 kHz.

The acquisition time is not fixed but it depends on the time each different model of WM needs to reach the centrifugal state. Nevertheless, the maximum time required for every test is less than 30 seconds, which makes it acceptable by industry and compatible with the production rate.

A typical measurement process consists of accelerating the motor and keeps the machine in the spinning phase for some seconds. During the measurement the machine is lifted up in order to guarantee vibration insulation from the production line. For the same reason a seismic mass has been used as basement for the laser support.

Once the signals have been acquired, the diagnostic system must be able to classify machines in two quality classes (good/faulty) according to their overall vibro-acoustic behaviour, for a pass or fail quality control. As additional information, the reason of the failure is sent to the main controller PC which manages the production line. In this way in the repairing station the operator can visualize the feature which exceeds the threshold, and link it to a particular fault (e.g. motor, pulley, etc.).

Signal Processing

The signal related to the RPM of the WM tub is calculated in real time by the LDV without needing of any additional device (e. g. photocell on the WM pulley or tachometer). The velocity signal is acquired for one second, it is analysed and the correspondent RPM value is calculated and displayed; then the signal is acquired for another second and analysed, and so on. The analysis is performed directly in the time domain, comparing the velocity signal with a reference sinusoidal signal and extracting the relative frequency.

Below the RPM signal calculated by the LDV:

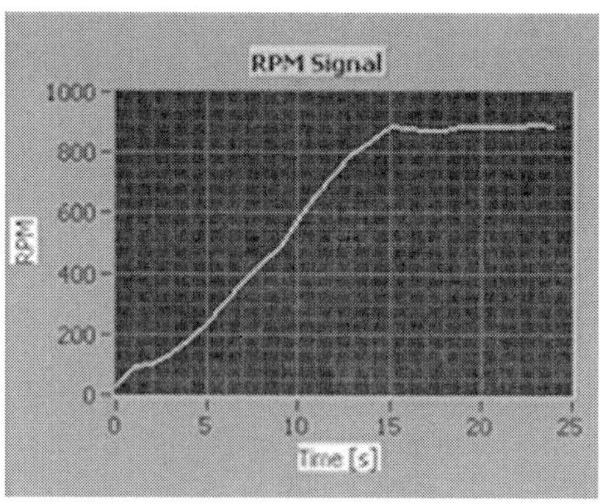

Figure 28. RPM signal related to the WM tub calculated in real time by the LDV.

The acquisition stops after some seconds (3 seconds are enough) the machine reaches the centrifugal phase specified by a fixed RPM value. In this way the system is completely automatic: once received the start for the acquisition, it acquires the velocity signal and in the meantime it calculates the RPM, it stops the acquisition, it analyses the signal and it gives out the final result OK/KO.

Due to the fact that the tub is moving in all directions and that the surface is not perfectly reflective, the acquired signal is affected by speckle noise [23] [24] [25]. It has been observed that the presence of the speckle noise depends on the optical level signal (a DC voltage proportional to the amount of backscattered light from the object under investigation). Figure 29 presents an amplitude spectrum comparison between two regions taken from a single velocity signal. The blue velocity spectrum corresponds to a quality signal with sufficient values, whereas the red one is distorted by speckle noise due to a low quality signal. As it can be seen, speckle noise resembles a wide-band noise which increases the overall level in the amplitude spectrum by almost 20 dB. This can mislead the diagnostic techniques and may cause wrong classifications.

Speckle noise reduction is achieved by the following steps:

- extraction of the sinusoidal signals corresponding to the shaft speed only;
- detection of the zero crossing of the sinusoidal signals;
- removal of period with critical speckle noise;
- connecting the remaining periods.

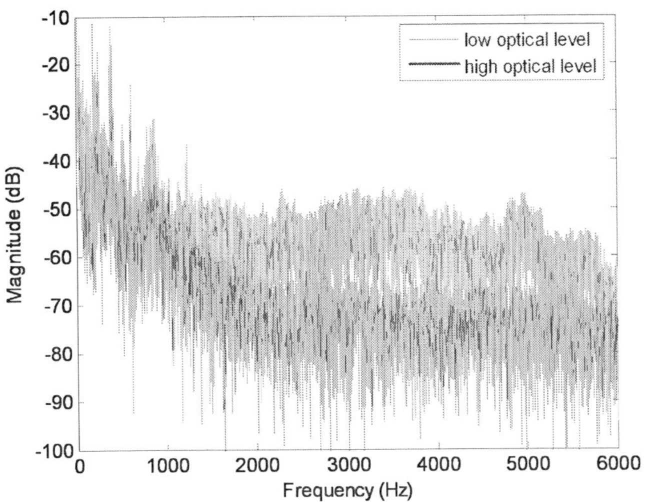

Figure 29. Spectrum comparison between two regions of the velocity signal measured with different quality levels: good (blue) and bad (red) quality.

Whenever the optical level is critically low, a new spike is generated and persists until a sufficient value of the optical level is reached, as shown in the figure below:

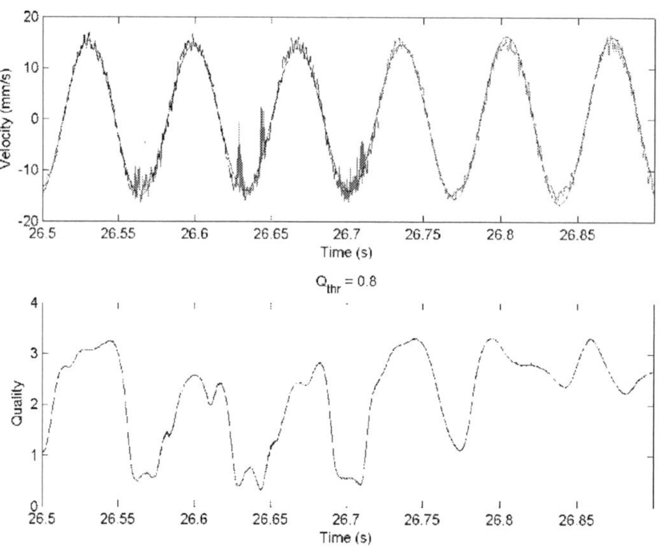

Figure 30. Successful selection of an undistorted region (green) and effect of low quality on the velocity signal (red).

Therefore the reduction of the speckle noise is based on thresholding the quality signal and removing all periods of the velocity signals which are distorted. The remaining periods are "joined" together using zero crossing in order to form a denoised velocity signal. This method cannot be applied on the run-up of the WM as the content is changing during the transient-state.

After denoising, the velocity signal is properly elaborated in the time domain, in the frequency domain and in the time-frequency domain. Some examples are illustrated in figures 31 and 32.

In order to allow the classification of good and faulty machines, the transient state has been analysed for 5 seconds and two features have been extracted. Seven additional features have been calculated by the analysis of the steady state in 3 seconds, both in the time and in the frequency domain.

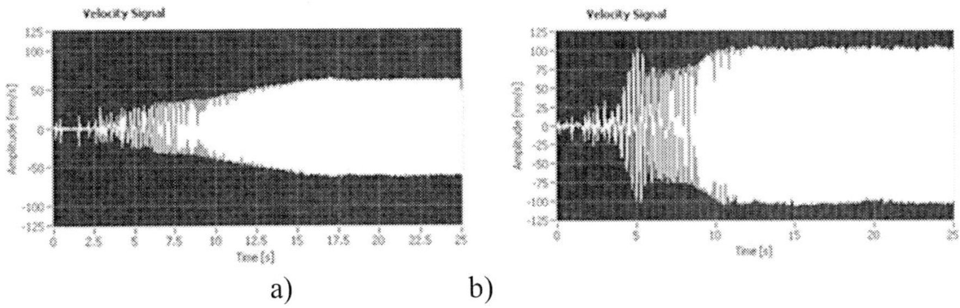

a) b)

Figure 31. Velocity time signals for a good (a) and faulty (b) WM.

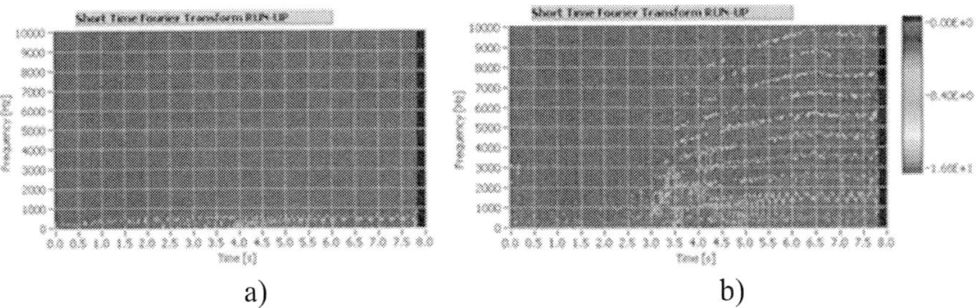

a) b)

Figure 32. STFT of velocity signals for a good (a) and faulty (b) WM during the run-up phase.

Feature Extraction and Classification

Most of the extracted features are related to the sum of the energy in specific bands.

Moreover, some bands and consequentially some features appear to be correlated to specific faults.

As follows an example related to a fault due to the motor.

As represented in Figure 33, the main frequency is related to the RPM of the Washing Machine. In fact, it is around 20 Hz which corresponds to the velocity of the tub (1200 RPM). The faulty machine shows additional frequencies around 280 Hz and 560 Hz. As it can be easily demonstrated, these are the frequencies related to the motor (the fundamental and the harmonic of the second order). In fact, according to the fact that the ratio between the RPM of the motor and WM RPM is 13.5, it follows: RPM motor = 13.5 x 1200 = 16.200 which means 270 Hz. In this case the features related to these frequency values are able to detect a defect related to the motor.

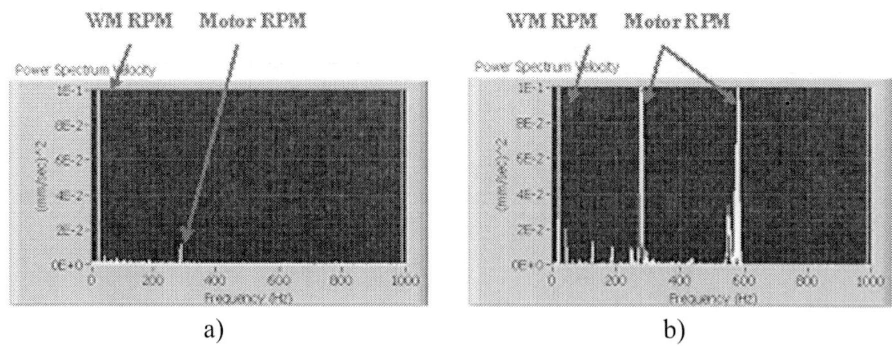

Figure 33. Power spectrum of the velocity time signals for a good (a) and faulty (b) WM.

Each feature is compared with the correspondent threshold. When at least one value is higher than its threshold, the machine is classified as faulty.

The software has been developed in the LabVIEW® programming language. An overview of the front panel of the Vibration Test System as it appears to the operator is reported.

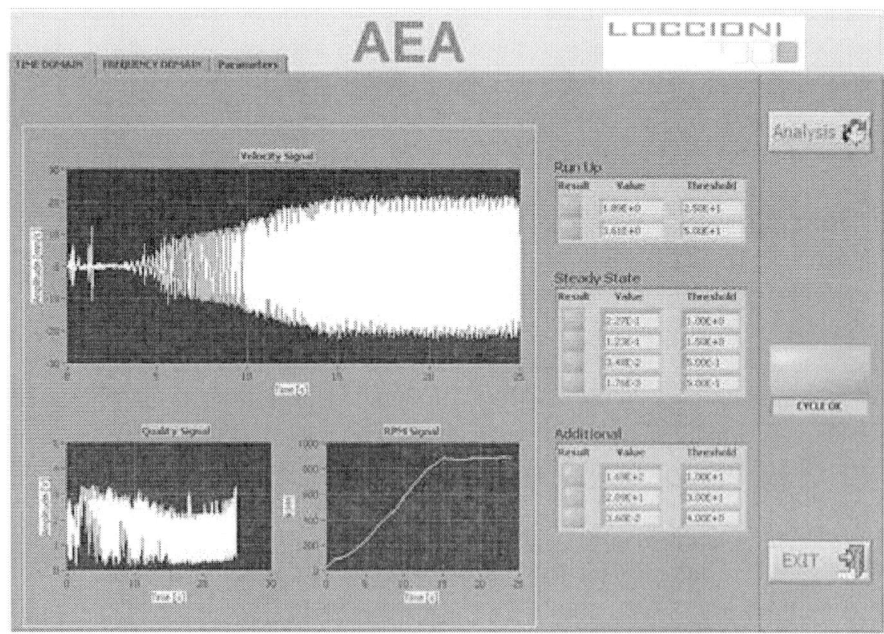

Figure 34. Front Panel of the Vibration test system for a good WM.

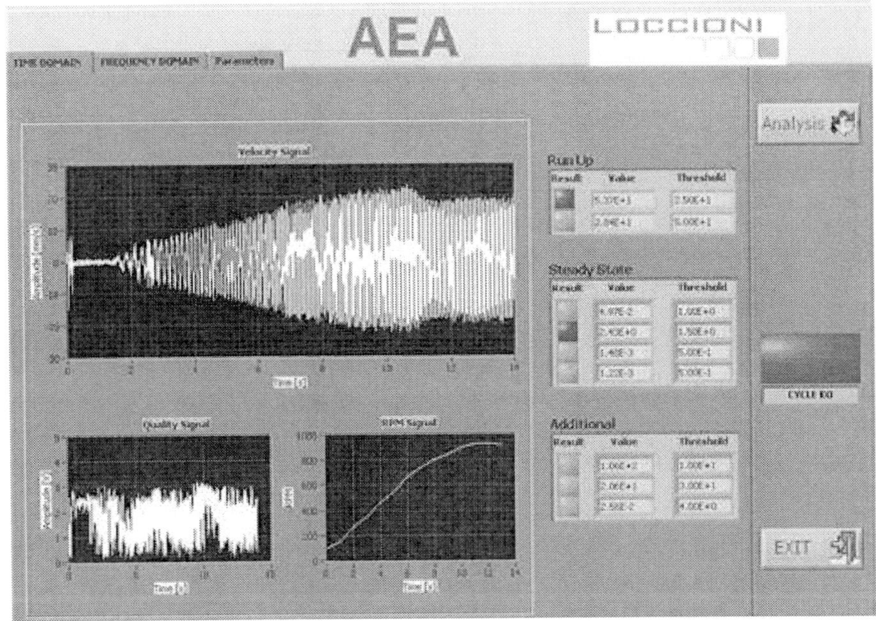

Figure 35. Front Panel of the Vibration test system for a faulty WM.

A statistical analysis has been performed during 7 months of production of two WM models (M1 and M2). The results on a total of 372139 washing machines measured by the vibration system is reported in Table 2.

Table 2. Classification results on 372139 WMs.

Model	Failure Number	Total Number	Percentage of Failure
M2	1486	302280	0.49%
M1	672	69859	0.96%

It can be observed that the percentage of defected machines detected by the vibration system is almost 0.5% of the total production for the model M2 and almost the 1% for the model M2.

Conclusion

The methods and results described in this chapter show how the developed Data Analysis System composed of appropriate sensors, data acquisition boards and Pattern Recognition algorithms can be successfully applied to mechanical defect diagnostics in the production lines. The method is applied both on signals acquired by accelerometer and LDV, on an electro-mechanical component such as the universal motor and on an assembled complex system such as the washing machine. Different signal processing techniques have been applied and peculiar features have been extracted in order to replace the subjectivity of human inspection testing with an objective assessment of product quality.

References

[1] E. O. Doebelin, *Measurement Systems - Application and Designs*. McGraw-Hill International Editions, Singapore, 4th edition, 1990.

[2] Brüel & Kjær, Vibration Transducers and Signal Conditioning. Brüel & Kjær Lecture Note. http://www.bksv.com/lectures/BA767512.pdf, 1998.

[3] PCB Piezotronics, *Microphones Handbook.*

[4] http://www.pcb.com/Linked_Documents/Vibration/Microphone_Handbook.pdf.

[5] Polytec GmbH, *Vibrometry Basics.* http://www.polytec.com/eur/158_942.asp.

[6] E.P. Tomasini, G.M. Revel and P. Castellini, Laser Based Measurement. *Encyclopedia of Vibration*, Pages 699-710. Academic Press, London, 2001.

[7] National Instruments™, *Data Acquisition (DAQ) Fundamentals.* Application Note 007.

[8] G.S. Paddan, Examples of a Measurement Artifact; the 'dc Shift'. *10th International Conference on Hand Arm Vibration*, Las Vegas, Nevada, USA, 7-11 June 2004.

[9] R. B. Randall, Diagnostics of rolling element bearings. *Machine condition monitoring*, Unit 6. Lecture notes.

[10] P. D. McFadden and J. D. Smith, Vibration monitoring of rolling element bearings by the high-frequency resonance technique - a review, *Tribology International* **17** (1), 3–10, 1984.

[11] N. Tandon and A. Choudhury, A review of vibration and acoustic measurement methods for the detection of defects in rolling element bearings. *Tribology International* **32** (8), 469–480, 1999.

[12] Brüel & Kjær, Software for PULSE X - Type 7700. *System Data.*

[13] D. G. Childers, D. P. Skinner and R. C. Kemerait, The cepstrum: a guide to processing. *Proceedings of the IEEE* **65** (10), 1428—1443, 1977.
[14] R. B. Randall, Cepstrum analysis and applications to gear diagnostics. *Machine condition monitoring*, Unit 7. Lecture notes.
[15] V. Oppenheim and R. W. Schafer, *Discrete-time signal processing*. Prentice Hall, Englewood Cliffs, New Jersey, 1989.
[16] National Instruments Corporation, *Order analysis toolset for LabVIEW user manual*. Part number 322879A-01, May 2001.
[17] Hewlett-Packard Company, *Effective machinery measurements using dynamic signal analyzers*. Application note 243-1, pub. no. 5962-7276E, 1994.
[18] S. Gade, H. Herlufsen, H. Konstantin-Hansen and N.J. Wismer, *Order tracking analysis*. Technical Review No. 2, Brüel & Kjær, 1995.
[19] K. R. Fyfe and E. D. S. Munck, Analysis of computed order tracking. *Mechanical Systems and Signal Processing* **11** (2), 187–205, 1997.
[20] K. M. Bossley, R. J. McKendrick, C. J. Harris and C. Mercer, Hybrid computer order tracking. *Mechanical Systems and Signal Processing* **13** (4), 627–641, 1999.
[21] National Instruments, *Signal Processing Toolset User Manual*. June 2001.
[22] J.J. Benetetto and M. Frazier, *Wavelets: Mathematics and Applications*. Studies in Advanced Mathematics, CRC Press, Boca Raton, Florida, 1992.
[23] R.M. Rodriguez, C. Cristalli and N. Paone, Comparative study between laser vibrometer and accelerometer measurements for mechanical fault detection of electric motor." 5th International Conference on Vibration Measurements by Laser Techniques: Advances and Applications, *SPIE Proceedings*, Vol. 4827, pp. 521-529, 2002.
[24] J. Vass and C. Cristalli, Bearing Fault Detection for On-line Quality Control of Electric Motors. *Proceedings of the 10th IMEKO TC10 International Conference on Technical Diagnostics,* Budapest, Hungary, pp. 93-97, ISBN 963 86586 4 9, June 2005.
[25] C. Cristalli, B. Torcianti and J. Vass, A new method for filtering speckle noise in vibration signals measured by Laser Doppler Vibrometry for on-line quality control. *Proc. SPIE* 6345, doi:10.1117/12.693104, 2006.
[26] Cristalli, B. Torcianti, J. Vass, P. Sovka and R.Šmíd, Speckle noise reduction in vibration signals measured by Laser Doppler Vibrometry, *Technical Article*, SPIE Newsroom, doi:10.1117/2.1200609.0390, http://newsroom.spie.org/x4735.xml, 2006.

INDEX

A

B

C

D

E

F

G

H

I

L

M

N

O

P